Project-Based Inquiry Science™

GOOD FRIENDS and GERMS

Janet L. Kolodner

Joseph S. Krajcik

Daniel C. Edelson

Brian J. Reiser

Mary L. Starr

IT's ABOUT TIME®

HERFF JONES EDUCATION DIVISION

84 Business Park Drive, Armonk, NY 10504
Phone (914) 273-2233 Fax (914) 273-2227
www.its-about-time.com

Program Components

Student Edition

Teacher's Planning Guide

Teacher's Resources Guide

Durable Equipment Kit

Consumable Equipment Kit

This project was supported, in part, by the **National Science Foundation**
under grant nos. 0137807, 0527341, 0639978.
Opinions expressed are those of the authors and not necessarily
those of the National Science Foundation.

PBIS Principal Investigators

Janet L. Kolodner is a Regents' Professor in the School of Interactive Computing in the Georgia Institute of Technology's College of Computing. Since 1978, her research has focused on learning from experience, both in computers and in people. She pioneered the Artificial Intelligence method called *case-based reasoning*, providing a way for computers to solve new problems based on their past experiences. Her book, *Case-Based Reasoning*, synthesizes work across the case-based reasoning research community from its inception to 1993.

Since 1994, Dr. Kolodner has focused on the applications and implications of case-based reasoning for education. In her approach to science education, called Learning by Design™ (LBD), students learn science while pursuing design challenges. Dr. Kolodner has investigated how to create a culture of collaboration and rigorous science talk in classrooms, how to use a project challenge to promote focus on science content, and how students learn and develop when classrooms function as learning communities. Currently, Dr. Kolodner is investigating how to help young people come to think of themselves as scientific reasoners. Dr. Kolodner's research results have been widely published, including in *Cognitive Science, Design Studies,* and the *Journal of the Learning Sciences.*

Dr. Kolodner was founding Director of Georgia Tech's EduTech Institute, served as coordinator of Georgia Tech's Cognitive Science program for many years, and is founding Editor in Chief of the *Journal of the Learning Sciences.* She is a founder of the International Society for the Learning Sciences, and she served as its first Executive Officer. She is a fellow of the American Association of Artificial Intelligence.

Joseph S. Krajcik is a Professor of Science Education and Associate Dean for Research in the School of Education at the University of Michigan. He works with teachers in science classrooms to bring about sustained change by creating classroom environments in which students find solutions to important intellectual questions that subsume essential curriculum standards and use learning technologies as productivity tools. He seeks to discover what students learn in such environments, as well as to explore and find solutions to challenges that teachers face in enacting such complex instruction. Dr. Krajcik has authored and co-authored over 100 manuscripts and makes frequent presentations at international, national, and regional conferences that focus on his research, as well as presentations that translate research findings into classroom practice. He is a fellow of the American Association for the Advancement of Science and served as president of the National Association for Research in Science Teaching. Dr. Krajcik co-directs the Center for Highly Interactive Classrooms, Curriculum and Computing in Education at the University of Michigan and is a co-principal investigator in the Center for Curriculum Materials in Science and The National Center for Learning and Teaching Nanoscale Science and Engineering. In 2002, Dr. Krajcik was honored to receive a Guest Professorship from Beijing Normal University in Beijing, China. In winter 2005, he was the Weston Visiting Professor of Science Education at the Weizmann Institute of Science in Rehovot, Israel.

Daniel C. Edelson is Vice President for Education and Children's Programs at the National Geographic Society. Previously, he was the director of the Geographic Data in Education (GEODE) Initiative at Northwestern University, where he led the development of Planetary Forecaster and Earth Systems and Processes. Since 1992, Dr. Edelson has directed a series of projects exploring the use of technology as a catalyst for reform in science education and has led the development of a number of software environments for education. These include My World GIS, a geographic information system for inquiry-based learning, and WorldWatcher, a data visualization and analysis system for gridded geographic data. Dr. Edelson is the author of the high school environmental science text, *Investigations in Environmental Science: A Case-Based Approach to the Study of Environmental Systems.* His research has been widely published, including in the *Journal of the Learning Sciences,* the *Journal of Research on Science Teaching, Science Educator,* and *The Science Teacher.*

Brian J. Reiser is a Professor of Learning Sciences in the School of Education and Social Policy at Northwestern University. Professor Reiser served as chair of Northwestern's Learning Sciences Ph.D. program from 1993, shortly after its inception, until 2001. His research focuses on the design and enactment of learning environments that support students' inquiry in science, including both science curriculum materials and scaffolded software tools. His research investigates the design of learning environments that scaffold scientific practices, including investigation, argumentation, and explanation; design principles for technology-infused curricula that engage students in inquiry projects; and the teaching practices that support student inquiry. Professor Reiser also directed BGuILE (Biology Guided Inquiry Learning Environments) to develop software tools for supporting middle school and high school students in analyzing data and constructing explanations with biological data. Reiser is a co-principal investigator in the NSF Center for Curriculum Materials in Science. He served as a member of the NRC panel authoring the report Taking Science to School.

Mary L. Starr is a Research Specialist in Science Education in the School of Education at the University of Michigan. She collaborates with teachers and students in elementary and middle school science classrooms around the United States who are implementing *Project-Based Inquiry Science.* Before joining the PBIS team, Dr. Starr created professional learning experiences in science, math, and technology, designed to assist teachers in successfully changing their classroom practices to promote student learning from coherent inquiry experiences. She has developed instructional materials in several STEM areas, including nanoscale science education, has presented at national and regional teacher education and educational research meetings, and has served in a leadership role in the Michigan Science Education Leadership Association. Dr. Starr has authored articles and book chapters, and has worked to improve elementary science teacher preparation through teaching science courses for pre-service teachers and acting as a consultant in elementary science teacher preparation. As part of the PBIS team, Dr. Starr has played a lead role in making units cohere as a curriculum, in developing the framework for PBIS Teacher Planning Guides, and in developing teacher professional development experiences and materials.

Acknowledgements

Three research teams contributed to the development of *Project-Based Inquiry Science* (PBIS): a team at the Georgia Institute of Technology headed by Janet L. Kolodner, a team at Northwestern University headed by Daniel Edelson and Brian Reiser, and a team at the University of Michigan headed by Joseph Krajcik and Ron Marx. Each of the PBIS units was originally developed by one of these teams and then later revised and edited to be a part of the full three-year middle-school curriculum that became PBIS.

PBIS has its roots in two educational approaches, Project-Based Science and Learning by Design™. Project-Based Science suggests that students should learn science through engaging in the same kinds of inquiry practices scientists use, in the context of scientific problems relevant to their lives and using tools authentic to science. Project-Based Science was originally conceived in the hi-ce Center at the University of Michigan, with funding from the National Science Foundation. Learning by Design™ derives from Problem-Based Learning and suggests sequencing, social practices, and reflective activities for promoting learning. It engages students in design practices, including the use of iteration and deliberate reflection. LBD was conceived at the Georgia Institute of Technology, with funding from the National Science Foundation, DARPA, and the McDonnell Foundation.

The development of the integrated PBIS curriculum was supported by the National Science Foundation under grants nos. 0137807, 0527341, and 0639978. Any opinions, findings and conclusions, or recommendations expressed in this material are those of the authors and do not necessarily reflect the views of the National Science Foundation.

PBIS Team

Principal Investigator
Janet L. Kolodner

Co-Principal Investigators
Daniel C. Edelson
Joseph S. Krajcik
Brian J. Reiser

NSF Program Officer
Gerhard Salinger

Curriculum Developers
Michael T. Ryan
Mary L. Starr

Teacher's Edition Developers
Rebecca M. Schneider
Mary L. Starr

Literacy Specialist
LeeAnn M. Sutherland

NSF Program Reviewer
Arthur Eisenkraft

Project Coordinator
Juliana Lancaster

External Evaluators
The Learning Partnership
Steven M. McGee
Jennifer Witers

The Georgia Institute of Technology Team

Project Director:
Janet L. Kolodner

Development of PBIS units at the Georgia Institute of Technology was conducted in conjunction with the Learning by Design™ Research group (LBD), Janet L. Kolodner, PI.

Lead Developers, Physical Science:
David Crismond
Michael T. Ryan

Lead Developer, Earth Science:
Paul J. Camp

Assessment and Evaluation:
Barbara Fasse
Jackie Gray
Daniel Hickey
Jennifer Holbrook
Laura Vandewiele

Project Pioneers:
JoAnne Collins
David Crismond
Joanna Fox
Alice Gertzman
Mark Guzdial
Cindy Hmelo-Silver
Douglas Holton
Roland Hubscher
N. Hari Narayanan
Wendy Newstetter
Valery Petrushin
Kathy Politis
Sadhana Puntambekar
David Rector
Janice Young

The Northwestern University Team

Project Directors:
Daniel Edelson
Brian Reiser

Lead Developer, Biology:
David Kanter

Lead Developers, Earth Science:
Jennifer Mundt Leimberer
Darlene Slusher

Development of PBIS units at Northwestern was conducted in conjunction with:

The Center for Learning Technologies in Urban Schools (LeTUS) at Northwestern, and the Chicago Public Schools
Clifton Burgess, PI
for Chicago Public Schools;
Louis Gomez, PI.

The BioQ Collaborative
David Kanter, PI.

The Biology Guided Inquiry Learning Environments (BGuILE) Project
Brian Reiser, PI.

The Geographic Data in Education (GEODE) Initiative
Daniel Edelson, Director

The Center for Curriculum Materials in Science at Northwestern
Daniel Edelson,
Brian Reiser,
Bruce Sherin, PIs.

The University of Michigan Team

Project Directors:
Joseph Krajcik
Ron Marx

Literacy Specialist:
LeeAnn M. Sutherland

Project Coordinator:
Mary L. Starr

Development of PBIS units at the University of Michigan was conducted in conjunction with:

The Center for Learning Technologies in Urban Schools (LeTUS)
Phyllis Blumenfeld,
Barry Fishman,
Joseph Krajcik,
Ron Marx,
Elliot Soloway, PIs.

The Detroit Public Schools
Juanita Clay-Chambers
Deborah Peek-Brown

The Center for Highly Interactive Computing in Education (hi-ce)
Phyllis Blumenfeld,
Barry Fishman,
Joseph Krajcik,
Ron Marx,
Elizabeth Moje,
Elliot Soloway,
LeeAnn Sutherland, PIs.

GF vi

Field-Test Teachers

National Field Test

Tamica Andrew
Leslie Baker
Jeanne Bayer
Gretchen Bryant
Boris Consuegra
Daun D'Aversa
Candi DiMauro
Kristie L. Divinski
Donna M. Dowd
Jason Fiorito
Lara Fish
Christine Gleason
Christine Hallerman
Terri L. Hart-Parker
Jennifer Hunn
Rhonda K. Hunter
Jessica Jones
Dawn Kuppersmith
Anthony F. Lawrence
Ann Novak
Rise Orsini
Tracy E. Parham
Cheryl Sgro-Ellis
Debra Tenenbaum
Sarah B. Topper
Becky Watts
Debra A. Williams
Ingrid M. Woolfolk
Ping-Jade Yang

New York City Field Test

Several sequences of PBIS units have been field- tested in New York City under the leadership of Whitney Lukens, Staff Developer for Region 9, and Greg Borman, Science Instructional Specialist, New York City Department of Education

6th Grade

Norman Agard
Tazinmudin Ali
Heather
   Guthartz Aniba
Asher Arzonane
Asli Aydin
Shareese Blakely
John J. Blaylock
Joshua Blum

Tsedey Bogale
Filomena Borrero
Zachary Brachio
Thelma Brown
Alicia Browne-Jones
Scott Bullis
Maximo Cabral
Lionel Callender
Matthew Carpenter
Ana Maria Castro
Diane Castro
Anne Chan
Ligia Chiorean
Boris Consuegra
Careen Halton Cooper
Cinnamon Czarnecki
Kristin Decker
Nancy Dejean
Gina DiCicco
Donna Dowd
Lizanne Espina
Joan Ferrato
Matt Finnerty
Jacqueline Flicker
Helen Fludd
Leigh Summers Frey
Helene Friedman-Hager
Diana Gering
Matthew Giles
Lucy Gill
Steven Gladden
Greg Grambo
Carrie Grodin-Vehling
Stephan Joanides
Kathryn Kadei
Paraskevi Karangunis
Cynthia Kerns
Martine Lalanne
Erin Lalor
Jennifer Lerman
Sara Lugert
Whitney Lukens
Dana Martorella
Christine Mazurek
Janine McGeown
Chevelle McKeever
Kevin Meyer
Jennifer Miller
Nicholas Miller
Diana Neligan
Caitlin Van Ness
Marlyn Orque
Eloisa Gelo Ortiz
Gina Papadopoulos
Tim Perez
Albertha Petrochilos

Christopher Poli
Kristina Rodriguez
Nadiesta Sanchez
Annette Schavez
Hilary Sedgwitch
Elissa Seto
Laura Shectman
Audrey Shmuel
Katherine Silva
Ragini Singhal
C. Nicole Smith
Gitangali Sohit
Justin Stein
Thomas Tapia
Eilish Walsh-Lennon
Lisa Wong
Brian Yanek
Cesar Yarleque
David Zaretsky
Colleen Zarinsky

7th Grade

Mayra Amaro
Emmanuel Anastasiou
Cheryl Barnhill
Bryce Cahn
Ligia Chiorean
Ben Colella
Boris Consuegra
Careen Halton Cooper
Elizabeth Derse
Urmilla Dhanraj
Gina DiCicco
Lydia Doubleday
Lizanne Espina
Matt Finnerty
Steven Gladden
Stephanie Goldberg
Nicholas Graham
Robert Hunter
Charlene Joseph
Ketlynne Joseph
Kimberly Kavazanjian
Christine Kennedy
Bakwah Kotung
Lisa Kraker
Anthony Lett
Herb Lippe
Jennifer Lopez
Jill Mastromarino
Kerry McKie
Christie Morgado
Patrick O'Connor
Agnes Ochiagha
Tim Perez
Nadia Piltser

Chris Poli
Carmelo Ruiz
Kim Sanders
Leslie Schiavone
Ileana Solla
Jacqueline Taylor
Purvi Vora
Ester Wiltz
Carla Yuille
Marcy Sexauer Zacchea
Lidan Zhou

8th Grade

Emmanuel Anastasio
Jennifer Applebaum
Marsha Armstrong
Jenine Barunas
Vito Cipolla
Kathy Critharis
Patrecia Davis
Alison Earle
Lizanne Espina
Matt Finnerty
Ursula Fokine
Kirsis Genao
Steven Gladden
Stephanie Goldberg
Peter Gooding
Matthew Herschfeld
Mike Horowitz
Charlene Jenkins
Ruben Jimenez
Ketlynne Joseph
Kimberly Kavazanjian
Lisa Kraker
Dora Kravitz
Anthony Lett
Emilie Lubis
George McCarthy
David Mckinney
Michael McMahon
Paul Melhado
Jen Miller
Christie Morgado
Ms. Oporto
Maria Jenny Pineda
Anastasia Plaunova
Carmelo Ruiz
Riza Sanchez
Kim Sanders
Maureen Stefanides
Dave Thompson
Matthew Ulmann
Maria Verosa
Tony Yaskulski

Good Friends and Germs

Good Friends and Germs is adapted from the unit, *How can good friends make me sick?* developed at University of Michigan as part of the work of the Center for Learning Technologies in Urban Schools and as a project of University of Michigan's Center for Highly Interactive Computing in Education. *Good Friends and Germs* was developed by the Project-Based Inquiry Science team in conjunction with the development team at the University of Michigan.

Project Directors:
Joseph Krajcik
Ron Marx

Lead Developer:
Barbara Hug

Other developers:
Amy Wefel

University of Michigan School of Public Health collaborators:
Toby Citrin
Renee Bayer

Detroit Public Schools Urban Systemic Initiative collaborator:
Deborah Peek¬-Brown

Pilot teachers
Yulonda Hale
Barbara Case
Susan Clark
Alicia Merriweather
Joy Reynolds

PBIS Team
Lead Developer:
Mary L. Starr

Contributing Field-test Teachers
Asher Arzonane
Suzy Bachman
Greg Borman
Matthew Carpenter
Anne Chan
Lizanne Espina
Enrique Garcia
Steven Gladden
Greg Grambo
Lillian Arlia Grippo
Dani Horowitz
Nicole Shiu Horowitz
Stephan Joanides
Verneda Johnson
Sunny Kam
Crystal Marsh
Kristin McNichol
Melissa Nathan
Tim Perez
Christopher Poli
Nadiesta Sanchez

Caitlin Van Ness
Melanie Wenger
Cesar Yarleque
Renee Zalewitz

The development of *Good Friends and Germs* was supported by National Science Foundation under grants no. 9553583, 9818828, and 0208059. The development of *How can good friends make me sick?* was supported by a grant from the Centers for Disease Control and Prevention and monitored by the Michigan Department of Community Health under contract no. DHHS-CDC-20030755 and by the National Science Foundation under grant no. 0830 310 A605. Any opinions, findings, and conclusions or recommendations expressed in this material are those of the authors and do not necessarily reflect the views of the CDC or the National Science Foundation.

Table of Contents

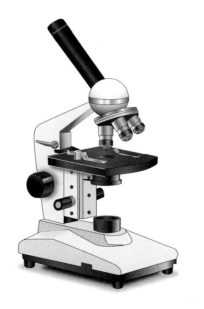

Learning Set 3

Science Concepts: *Tissues, organs, organ systems, breathing rate and heart rate, respiratory system, circulatory system, digestive system, immune system, immunity, lymphatic system, inflammation, nervous system, endocrine system, reproductive system, excretory system, skeletal system, muscular system. use of evidence, recommendation, modeling and simulation, explanation.*

Learning Set 4

Science Concepts: *Disease outbreak, virus mutation, disease containment, tracking disease, epidemic, explanation, use evidence, recommendation.*

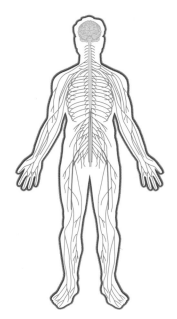

Introducing PBIS

What Do Scientists Do?

1) Scientists...address big challenges and big questions.

You will find many different kinds of *Big Challenges* and *Questions* in PBIS Units. Some ask you to think about why something is a certain way. Some ask you to think about what causes something to change. Some challenge you to design a solution to a problem. Most of them are about things that can and do happen in the real world.

Understand the Big Challenge or Question

As you get started with each Unit, you will do activities that help you understand the *Big Question* or *Challenge* for that Unit. You will think about what you already know that might help you, and you will identify some of the new things you will need to learn.

Project Board

The *Project Board* helps you keep track of your learning. For each challenge or question, you will use a *Project Board* to keep track of what you know, what you need to learn, and what you are learning. As you learn and gather evidence, you will record that on the *Project Board*. After you have answered each small question or challenge, you will return to the *Project Board* to record how what you've learned helps you answer the *Big Question* or *Challenge*.

Learning Sets

Each Unit is composed of a group of *Learning Sets*, one for each of the smaller questions that need to be answered to address the *Big Question* or *Challenge*. In each *Learning Set*, you will investigate and read to find answers to the *Learning Set's* question. You will also have a chance to share the results of your investigations with your classmates and work together to make sense of what you are learning. As you come to understand answers to the questions on the *Project Board*, you will record those answers and the evidence you've collected. At the end of each *Learning Set*, you will apply your knowledge to the *Big Question* or *Challenge*.

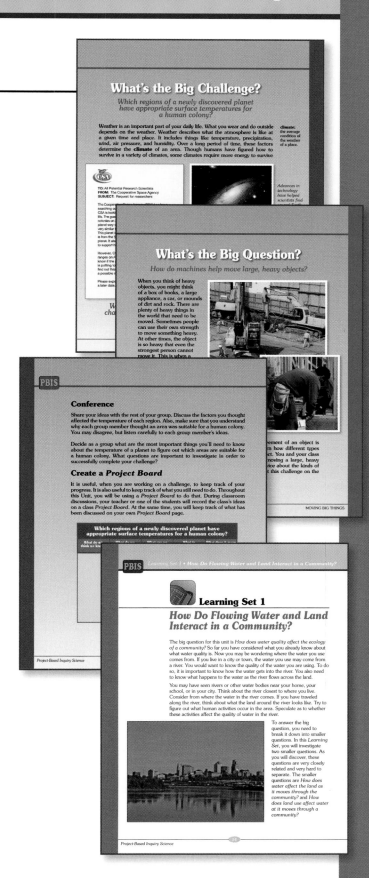

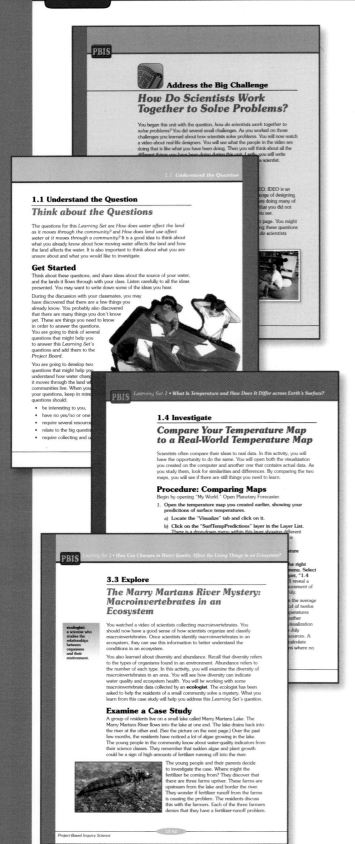

Answer the Big Question / Address the Big Challenge

At the end of each *Unit*, you will put everything you have learned together to tackle the *Big Question or Challenge*.

2) Scientists...address smaller questions and challenges.

What You Do in a Learning Set

Understanding the Question or Challenge

At the start of each *Learning Set*, you will usually do activities that will help you understand the *Learning Set's* question or challenge and recognize what you already know that can help you answer the question or achieve the challenge. Usually, you will visit the *Project Board* after these activities and record on it the even smaller questions that you need to investigate to answer a *Learning Set's* question.

Investigate/Explore

There are many different kinds of investigations you might do to find answers to questions. In the *Learning Sets,* you might

- design and run experiments;

- design and run simulations;

- design and build models;

- examine large sets of data.

Don't worry if you haven't done these things before. The text will provide you with lots of help in designing your investigations and in analyzing your data.

Project-Based Inquiry Science

Read

Like scientists, you will also read about the science you are learning. You'll read a little bit before you investigate, but most of the reading you do will be to help you understand what you've experienced or seen in an investigation. Each time you read, the text will include *Stop and Think* questions after the reading. These questions will help you gauge how well you understand what you have read. Usually, the class will discuss the answers to *Stop and Think* questions before going on so that everybody has a chance to make sense of the reading.

Design and Build

When the *Big Challenge* for a Unit asks you to design something, the challenge in a *Learning Set* might also ask you to design something and make it work. Often, you will design a part of the thing you will design and build for the *Big Challenge*. When a *Learning Set* challenges you to design and build something, you will do several things:

- identify what questions you need to answer to be successful

- investigate to find answers to those questions

- use those answers to plan a good design solution

- build and test your design.

Because designs don't always work the way you want them to, you will usually do a design challenge more than once. Each time through, you will test your design. If your design doesn't work as well as you'd like, you will determine why it is not working and identify other things you need to investigate to make it work better. Then, you will learn those things and try again.

Explain and Recommend

A big part of what scientists do is explain, or try to make sense of why things happen the way they do. An explanation describes why something is the way it is or behaves the way it does. An explanation is a statement you make built from claims (what you think you know), evidence (from an investigation) that supports the claim, and science knowledge. As they learn, scientists get better at explaining. You'll see that you get better, too, as you work through the *Learning Sets*.

A recommendation is a special kind of claim—one where you advise somebody about what to do. You will make recommendations and support them with evidence, science knowledge, and explanations.

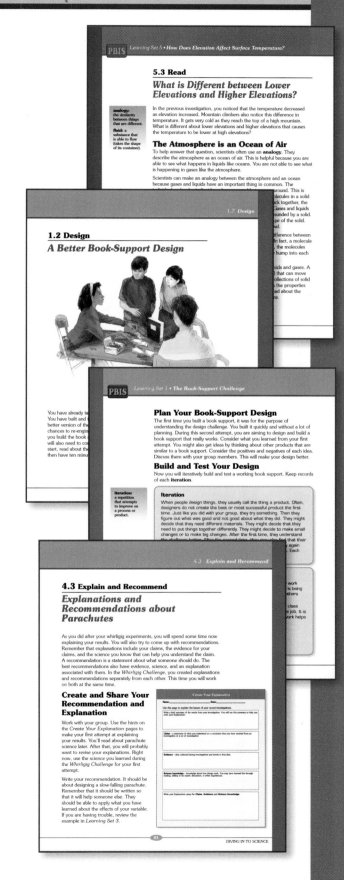

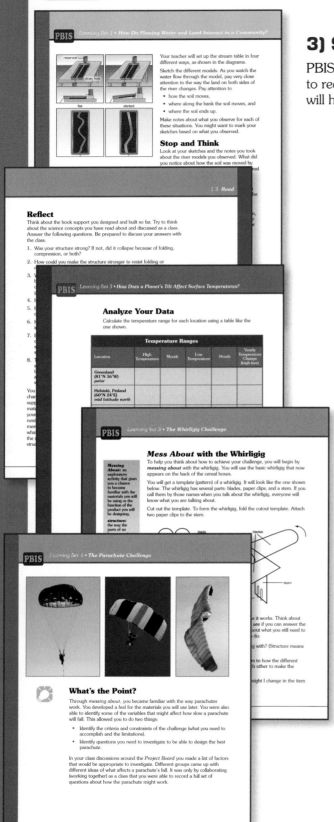

3) Scientists...reflect in many different ways.

PBIS provides guidance to help you think about what you are doing and to recognize what you are learning. Doing this often as you are working will help you be a successful student scientist.

Tools for Making Sense

Stop and Think

Stop and Think sections help you make sense of what you've been doing in the section you are working on. *Stop and Think* sections include a set of questions to help you understand what you've just read or done. Sometimes the questions will remind you of something you need to pay more attention to. Sometimes they will help you connect what you've just read to things you already know. When there is a *Stop and Think* in the text, you will work individually or with a partner to answer the questions, and then the whole class will discuss the answers.

Reflect

Reflect sections help you connect what you've just done with other things you've read or done earlier in the Unit (or in another Unit). When there is a *Reflect* in the text, you will work individually, with a partner or your small group to answer the questions. Then, the whole class will discuss the answers. You may be asked to answer *Reflect* questions for homework.

Analyze Your Data

Whenever you have to analyze data, the text will provide hints about how to do that and what to look for.

Mess About

"Messing about" is a term that comes from design. It means exploring the materials you will be using for designing or building something or examining something that works like what you will be designing. Messing about helps you discover new ideas—and it can be a lot of fun. The text will usually give you ideas about things to notice as you are messing about.

What's the Point?

At the end of each *Learning Set*, you will find a summary, called *What's the Point?*, of the important information from the *Learning Set*. These summaries can help you remember how what you did and learned is connected to the *Big Question or Challenge* you are working on.

4) Scientists...collaborate.

Scientists never do all their work alone. They work with other scientists (collaborate) and share their knowledge. PBIS helps you be a student scientist by giving you lots of opportunities for sharing your findings, ideas, and discoveries with others (the way scientists do). You will work together in small groups to investigate, design, explain, and do other things. Sometimes you will work in pairs to figure out things together. You will also have lots of opportunities to share your findings with the rest of your classmates and make sense together of what you are learning.

Investigation Expo

In an *Investigation Expo*, small groups report to the class about an investigation they've done. For each *Investigation Expo*, you will make a poster detailing what you were trying to learn from your investigation, what you did, your data, and your interpretation of your data. The text gives you hints about what to present and what to look for in other groups' presentations. *Investigation Expos* are always followed by discussions about the investigations and about how to do science well. You may also be asked to write a lab report following an investigation.

Plan Briefing/Solution Briefing/Idea Briefing

Briefings are presentations of work in progress. They give you a chance to get advice from your classmates that can help you move forward. During a *Plan Briefing*, you present your plan to the class. It might be a plan for an experiment or a plan for solving a problem or achieving a challenge. During a *Solution Briefing*, you present your solution in progress and ask the class to help you make your solution better. During an *Idea Briefing*, you present your ideas. You get the best advice from your classmates when you present evidence in support of your plan, solution, or idea. Often, you will prepare a poster to help you make your presentation. Briefings are almost always followed by discussions of your investigations and how you will move forward.

Solution Showcase

Solution Showcases usually appear near the end of a Unit. During a *Solution Showcase*, you show your classmates your finished product—either your answer to a question or your solution to a challenge. You also tell the class why you think it is a good answer or solution, what evidence and science you used to get to your solution, and what you tried along the way before getting to your answer or solution. Sometimes a *Solution Showcase* is followed by a competition. It is almost always followed by a discussion comparing and contrasting the different answers and solutions groups have come up with. You may be asked to write a report or paper following a *Solution Showcase*.

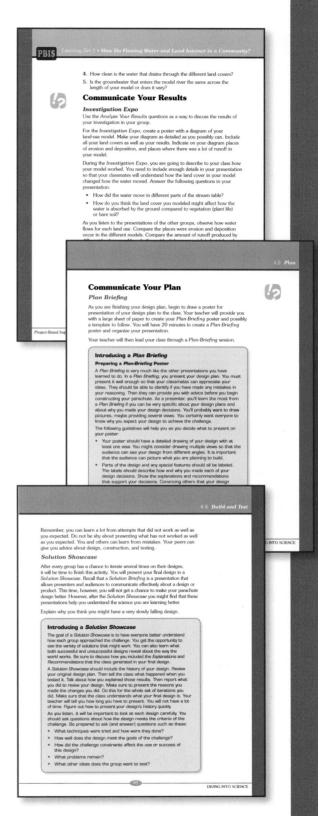

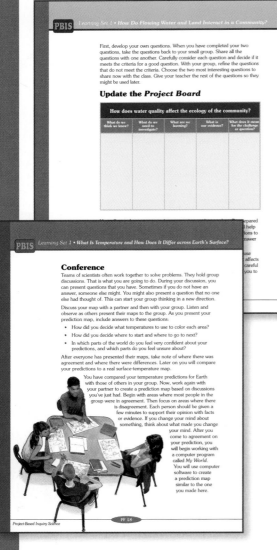

Update the Project Board

Remember that the *Project Board* is designed to help the class keep track of what they are learning and their progress towards a Unit's *Big Question* or *Challenge*. At the beginning of each Unit, the class creates a *Project Board*, and together you record what you think you know about answering the *Big Question* or addressing the *Big Challenge* and what you think you need to investigate further. Near the beginning of each *Learning Set*, the class revisits the *Project Board* and adds new questions and things they think they know. At the end of each *Learning Set*, the class again revisits the *Project Board*. This time you record what you have learned, the evidence you've collected, and recommendations you can make about answering the *Big Question* or achieving the *Big Challenge*.

Conference

A *Conference* is a short discussion between a small group of students before a more formal whole-class discussion. Students might discuss predictions and observations, they might try to explain together, they might consult on what they think they know, and so on. Usually, a *Conference* is followed by a discussion around the *Project Board*. In these small group discussions, everybody gets a chance to participate.

 What's the Point? Review what you have learned in each *Learning Set*.

 Communicate Share your ideas and results with your classmates.

 Stop and Think Answer questions that help you understand what you've done in a section.

 Record Record your data as you gather it.

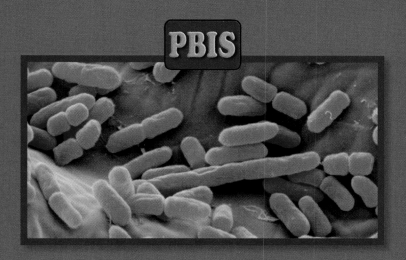

GOOD FRIENDS AND GERMS

As a student scientist, you will...

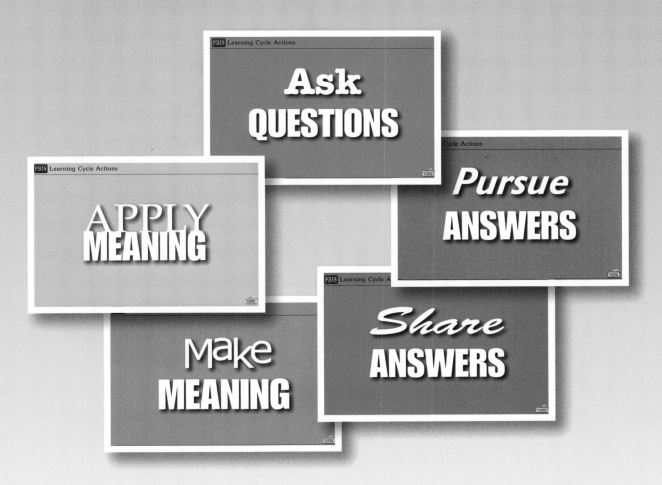

What's the Big Question?

How can you prevent your good friends from getting sick?

This year you may get five or six colds. Colds make you feel sick. You might get a stuffy nose, watery eyes, and a cough. How did you get that cold? What causes you to be sick when you have a cold or the flu? What changes take place in your body? These are some of the questions you might ask as part of this science Unit.

Why is it important to be able to answer these questions? You probably have heard people say that the flu or another illness is "going around." If you know how the flu spreads, you can try to protect yourself from getting sick. The more you know, the healthier you can keep your own body.

Look at the *Big Question* of this science Unit: *How can you prevent your good friends from getting sick?* All of the investigations and readings you will be doing during the next few weeks will help you answer this question.

It is a very big question. You will be able to answer the question more easily if you break it up into smaller questions. You can use the smaller questions to begin investigations and read about science concepts. As you answer each of these smaller questions, you will be solving pieces of a puzzle that will help you answer the *Big Question*. You will combine everything you learn in answering smaller questions to answer the *Big Question* at the end of the Unit. In doing so, you will be doing what scientists do. During this Unit, you will also use many of the science practices you learned about and used in other Units.

Welcome to *Good Friends and Germs.*
Enjoy your journey as a student scientist.

Think about the Big Question

Before you start, it's a good idea to make sure you understand what you know about the *Big Question*. What do you think you know about preventing your friends from getting sick? What are you not sure about?

Get Started

You are going to share what you know about how you get sick. You will pass a beanbag around the classroom. When you receive the beanbag, hold onto it and say aloud one thing you know about how you get sick. Do not worry if others have already mentioned your idea. Listen carefully to what your classmates have to say when they receive the beanbag. You may wish to record some of the ideas you hear.

After you complete the beanbag-sharing activity, use an ultraviolet flashlight to look around the classroom. What do you see? Do you notice a glow around the classroom? A powder that glows under ultraviolet light was put on the beanbag. As you passed the bag around, all of you touched the powder. When you touched different places with powder on your hands, the powder rubbed off onto the surfaces.

Stop and Think

1. Which surfaces were touched the most in your classroom?

2. Which parts of your hands and head have powder on them?

3. What is the farthest point in your classroom that the powder was moved?

4. What surprised you about where you found the powder?

Germs Can Be Anywhere

The powder on the beanbag is like germs. When you get germs on your hands, you move them from one place to another, just like you moved the powder. You touch your face, the doorknob, your pencil, and your friends. Like the powder, the germs move from your hands to your face, your friends' hands, and all over anything you touch. When you put your hands on your face, you move the germs there. The germs on your face, especially near your eyes, nose, or mouth can then get inside your body.

Could you tell that the beanbag, your hands, and other people's hands had powder on them before you started passing the beanbag? You probably did not know the beanbag was covered with powder because the powder is very hard to see. It is not possible to see germs with your eyes either.

Germs can be on the things you touch. Doorknobs, handles, chairs, beanbags, and many other things can be covered with germs. When you touch these things, germs can be moved from the object to your hands, just like the glowing powder moved from the beanbag to your hands, and then from your hands to other surfaces.

Wash your hands thoroughly to remove all the powder.

Procedure

Remove the powder from your hands. Wash your hands, and then use the light to see if you have removed all the powder. Look especially around your fingernails and in the creases of your skin on the palms of your hands. How well did you wash your hands? If you can still see powder on your hands, wash them again.

Your Challenge

In this Unit, you will need to respond to a challenge to answer the *Big Question, How can you prevent your good friends from getting sick?* The challenge has two parts. In the first part, the class will study several diseases. Your group will study one disease. You will describe the disease and how you know somebody has it. You will report about how the disease affects the body. You will explain how it is spread and how it is treated. You will then make recommendations about how to prevent spreading the disease to others. Other groups will study different diseases. Each group will report to the class about their disease.

In the second part of the challenge, you will make recommendations about how you can stay healthy and prevent others from getting sick. Then, as a class, you will choose recommendations to present to other students.

Conference

Remember that the *Big Question* for this Unit is *How can you prevent your good friends from getting sick?* You need to break down this question into smaller questions that you can investigate. As a group, review what you and your classmates said you thought you knew. You probably have a lot of questions that need to be answered before you can fully answer the *Big Question*. The powder activity may have reminded you of some things that you think you know, or don't know, about germs. Record these things so that you will remember what you discussed when you share again with the class. Which questions that you would like answered are important for answering the *Big Question?*

Create a *Project Board*

When you work on a project, it is useful to keep track of your progress and what you still need to do. You will use a *Project Board* to do that. It is designed to help your class organize its questions, investigations, results, and conclusions. The *Project Board* will also help you to decide what you are going to do next. During classroom discussions, you will record the class's ideas on a class *Project Board*. At the same time, you will keep track of what has been discussed on your own *Project Board* page.

The *Project Board* has space for answering five guiding questions:

- What do we think we know?
- What do we need to investigate?
- What are we learning?
- What is our evidence?
- What does it mean for the challenge or question?

How can you prevent good friends from getting sick?				
What do we think we know?	What do we need to investigate?	What are we learning?	What is our evidence?	What does it mean for the challenge or question?

To begin, write the *Big Question, How can you prevent your good friends from getting sick?* on the *Project Board*.

The activity you just completed was meant to help you recall what you think you know about how you can prevent your good friends from getting sick. Discuss the things you and your classmates think you know about how you get sick and what happens when you get sick. Record these things in the first column of the *Project Board*.

In the second column, record the things you need to investigate to answer the *Big Question*. During your discussions, you and others in your class may have disagreed about some ideas. This second column is designed to help you keep track of things that are debatable or that you do not know enough about yet.

What's the Point?

Often, people try to solve problems without first understanding what they need to do. You have just spent some time thinking about what you need to learn to answer the *Big Question*.

In the activity, you were probably surprised by how easily the powder spread. The powder is similar to germs that spread disease. Understanding how germs spread will help you understand why you get sick.

You started a *Project Board* to help track what you understand. You also added questions about what things you need to learn to prevent your friends from getting sick. The *Project Board* is a space to help the class work together to understand and solve problems. Using it will help you have good science discussions as you work on a project.

Learning Set 1

How Do You Get Sick?

The *Big Question* that you have to answer in this Unit is *How can you prevent your good friends from getting sick?* This is a very big question. To answer this *Big Question*, you need to break it down into smaller questions. There are many smaller questions you might need to ask. You already recorded some of these questions on the *Project Board*.

Think about the last time you were sick. Were any of your friends sick at the same time? Did you get sick after someone in your class or family was sick? The smaller question you will be looking at in this *Learning Set* is *How do you get sick?*

1.1 Understand the Question

Thinking about How You Get Sick

The question for this *Learning Set* is *How do you get sick?* It is a good idea to think about what you already know about how you get sick. It is also important to think about what you are unsure about and what you would like to investigate.

Get Started

You thought about how you get sick during the glow-powder activity. Share any additional ideas you have with your group. Listen carefully to all the ideas presented. You may want to record some of the ideas you hear. Then on your own, think of several questions that might help you answer the question for this *Learning Set*. Develop two questions that might help you understand how you get sick. When you write your questions, keep in mind that your questions should

- be interesting to you;

- require several resources to answer;

- relate to the *Big Question*; and

- require collecting and using data.

Your questions should not have yes/no or one-word answers.

When you have completed your questions, take the questions back to your small group. Share all the questions with each other. Carefully consider each question and decide if it meets the criteria for a good question. With your group, refine the questions that do not meet the criteria. Choose the two most interesting questions to share with the class. Give your teacher the rest of the questions so they might be used later.

Update the *Project Board*

You will now share your questions with your class. Be prepared to support your questions with the above criteria. Your teacher will help you with the criteria if needed. Then, your teacher will add your questions to the *Project Board*. In this *Learning Set,* you will work to answer some of these questions.

1.2 Investigate

Simulate the Spread of Disease

To begin to answer the question, *How do you get sick?*, you will run a simulation. You will **simulate** what could happen when you make contact with others. This investigation is going to require a lot of observing.

Be a Scientist

Simulations

At the beginning of this Unit, you passed a beanbag around the classroom. There was glow powder on the beanbag. The glow powder passed from one person to another and around the room. This activity was used to show you how germs might spread. The powder was used because it would not make sense to use real germs that could make you sick. This type of activity is called a **simulation**. Simulations imitate, or act out, what happens in real life. Scientists often use simulations. They use them when what they want to study is too big, too small, too fast, too slow, or too dangerous to investigate directly. The glow-powder activity simulated how disease might spread from person to person, but did so without causing any harm or danger to anyone.

---

simulation: the process or act of imitation or acting out.

simulate: to imitate how something happens in a real-world situation by acting it out with a model.

Materials

- clear plastic cups containing a liquid

- plastic pipettes

Procedure

1. Get a cup with some liquid in it and a pipette. Fill your pipette half full of liquid.

2. Walk around the room and introduce yourself to your classmates (friends). As you do this, put the liquid in your pipette into your friend's cup. Your friend will also share the liquid in his/her pipette with you.

3. Continue to meet your classmates until time is up. It is O.K. to meet the same person twice.

4. After you have interacted with several classmates, your teacher will test your liquid. If your liquid changes color, you are sick.

Do not drink the liquid. It may not be water.

Be careful not to spill any liquid. Wipe up any spills immediately.

5. Based on your observations, answer these questions.

a) What happened to your liquid?

b) Were you infected, or did the disease miss you?

initial carrier:
the first person in
a group to get a
contagious disease.

c) Who do you think was the **initial carrier**?

Recording Your Data

In this activity, you and your classmates interacted with each other and shared your liquids. The interaction between you and your classmates was like meeting each other in a real-life situation. One of the liquids was "sick" (infected).

It might be easier to identify the initial carrier if you record with whom you came in contact. Use a table similar to the one shown below. Write the names of the people you interacted with first, then second, then third, all the way to the last person.

Liquid Interactions Table			1.2

Name: _____ Date: _____

Interaction	Name	Time of Interaction
1		
2		
3		
4		
5		
6		
7		
8		
9		
10		

Reflect

Answer the following questions. Be prepared to discuss your answers with the class.

1. How easy was it for you to fill in the data table?

2. How reliable do you think your data is?

3. Can you identify the initial carrier now?

4. If you were to repeat the activity, what could you do to improve your data collection?

Procedure: Repeat the Simulation

You may have decided that it would have been easier to record your data while you were having your interactions. Repeat the simulation. This time record your data as you do the activity.

Compile your class data. As a class, try to figure out who was the initial carrier.

Stop and Think

Answer the following questions. Be prepared to discuss your answers with the class.

1. It is possible to begin finding the initial carrier by looking at whose cups are not infected. How can you eliminate others?

2. Who do you think was the initial carrier? Why? What made the person easier to find this time?

3. Which cup(s) can you rule out as being the one that first had the disease? How do you know?

4. Now that you have finished the activity, how would you explain how a disease spreads through a community?

5. This investigation was a simulation. How do you think this investigation is like spreading a real disease? How do you think it is different from what happens in real life?

Be a Scientist

Models and Simulations

Simulations use a **model** to imitate, or act out, real-life situations.

A model is a **representation** of something in the world. One model that you know is a globe. The parts of the globe represent parts of the Earth. Scientists use models to investigate things that are too difficult or too dangerous to examine in real life. The models are at a size that people can easily examine. To use a model to investigate, the model needs to be similar to the real world in ways that are important for what the scientist is investigating.

Sometimes what you want to model is a situation or an event. To do this, you create a model that includes the things that are part of an event and then you use that to act out a situation in a simulation.

In the glow-powder activity, you simulated how disease spreads by using a model of germs and things in the world that might have germs on them. The glow dust corresponded to germs. The beanbag corresponded to things in the world that can have germs on them. As you passed around the beanbag with glow powder on it and touched other things in the classroom, you simulated how germs can be spread from place to place. That simulation allowed you to experience how easily germs spread.

You also participated in an activity that simulated how diseases are transmitted as you meet friends. Each cup with water represented a person. The chemical in the water of one of the cups represented a disease. Each time you interacted with somebody, you passed some of your water to him or her. That person passed some water to you. Your water became "infected" after you interacted with someone with "infected" water. After your water was "infected," you "infected" the water of everybody you interacted with. You simulated how diseases are transmitted using a model that included people with a disease and without a disease.

Models and simulations help scientists learn. The simulations you did helped you learn about how germs spread and how diseases are transmitted. You will experience several other models in this Unit and throughout PBIS.

model: a way of representing something in the world to learn more about it.

representation: a likeness or image of something.

Molecular model.

Epidemiology

Epidemiology is the study of what affects health and illness. (The word comes from Greek words that mean "among the people.") **Epidemiologists** are scientists who study diseases and how they spread. Their work is very similar to what you did. They look at who has a disease and who does not and try to find connections. When they are studying the spread of disease, they look for the person who begins the spread of the disease in a group. They call this person the "initial carrier" or "**sentinel case**." The first person then gives the disease to someone else through an interaction. Epidemiologists would call this first interaction the "initial step."

epidemiology: the study of how and why diseases occur.

epidemiologist: a scientist who studies diseases and how they spread.

sentinel case: a person who begins the spread of disease in a group.

What's the Point?

This activity, like the one you did earlier, was a simulation. Simulations are important in science when you cannot use or do the real thing. You cannot make people sick to test how diseases spread.

When you repeated the simulation, you recorded your interactions as they happened. You found it much easier to track the disease. This shows you how important it is to keep good records.

As you interacted with more and more people, you may have caught and then spread the disease. The disease may have infected each member of your class. This simulation demonstrates some important points about how diseases spread. It showed that sometimes you cannot know who is sick. Often, sick people do not appear sick and yet they can spread their disease. The investigation also demonstrates that spreading diseases can be very easy. However, it is not always easy to figure out who the initial carrier was.

1.3 Read

What Are Communicable Diseases?

communicable
disease: a disease
that spreads very
readily from person
to person.

non-
communicable
disease: a disease
that cannot be
passed on by the
person who is sick
to other people.

Diseases that spread from one person to another are called **communicable diseases**. You probably know a lot of communicable diseases, like colds and the flu. But only some diseases are communicable.

Imagine a visitor comes into your house and is sneezing and coughing. You might think the visitor has a cold. If he or she does, you can get the cold from them. The simulation you just did showed you that. A cold is a disease one person can give to another person. As you learn more about diseases, you will want to think about which ones are communicable and which ones are not. Only the communicable ones can spread from one person to another. Diseases that are not communicable cannot do that.

Cancer is one example of a **noncommunicable disease**. If a visitor to your house has cancer, you cannot get cancer from him or her. It does not matter if both of you drink from the same glass or if that person sneezes into the air. You cannot "catch" cancer. You may know someone who has a noncommunicable disease like cancer. These diseases make you sick, but they cannot spread from one person to another. Some other diseases that are not communicable are diabetes, heart disease, high blood pressure, and asthma.

Communicable diseases, such as colds, can spread from one person to another.

Stop and Think

1. How is a communicable disease different from a noncommunicable disease?

2. What are two communicable diseases?

3. What are two noncommunicable diseases?

Explain

Make your first attempt at explaining how you can get sick with communicable diseases. Develop a claim about how you can get sick with communicable diseases, and record it on a *Create Your Explanation* page. Then use observations from the simulations you've run as evidence to support your claim. Write a statement using your evidence and science knowledge that supports your claim. This will be your explanation. You may find it hard to develop an explanation right now. Do your best with what you know now. You will have chances later in the Unit to revise your claims and explanations.

Recommend

Now, make your first recommendations about staying healthy. Develop recommendations that answer two questions:

- How can I keep from getting sick with a communicable disease?

- How can I prevent my friends from getting sick with a communicable disease?

Record each of your recommendations as a claim on a *Create your Explanation* page. Record evidence and science knowledge that support your recommendation. Then write a statement that shows why your recommendation is trustworthy. You will have chances later in the Unit to revise your recommendations and to better explain them.

Learning Set 1

Back to the Big Question

How can you prevent your good friends from getting sick?

The *Big Question* for this Unit is *How can you prevent your good friends from getting sick?* You began this Unit by investigating with your class how diseases spread. You used simulations to find out how you can get sick. You found that even one of your good friends can make you sick.

Epidemiologists spend a lot of time studying how diseases spread. By repeating a simulation, you learned how important it is for scientists and for you to accurately record data. Keeping accurate records of your work will be important as you work toward answering the *Big Question*.

You also read about the difference between communicable and non-communicable diseases. Communicable diseases spread from person to person. There are many communicable diseases you already know about and others you will learn about. For the next few weeks, you will study different diseases and learn more about how they can spread from one person to another.

Reflect

Think about and answer the following questions. Be prepared to discuss your answers with your class.

1. What did you learn about the importance of keeping accurate records? Use an example from this *Learning Set* to justify your answer.

2. What is the difference between a communicable and a noncommunicable disease?

3. How do you get sick? Use evidence from this Learning Set to support your answer.

4. What is a model? Why are models important in science?

5. How did you use models to run simulations in this *Learning Set*?

Update the *Project Board*

You started a *Project Board* centered on the idea of learning more about how you can prevent your good friends from getting sick. Now you have done a simulation. You know more about the factors that affect how a disease spreads. You are now ready to fill in the *Project Board* more completely.

Up to now, you only recorded information in the first two columns of the *Project Board*. You will now focus on the next two columns. These are the *What are we learning?* and *What is our evidence?* columns. When you record what you are learning in the third column, you will be answering some questions in the *What do we need to investigate?* column. You will describe what you learned from the investigation you just did. But you cannot just write what you learned without providing evidence for your conclusions. Evidence is necessary to answer scientific questions. You will fill in the evidence column based on data and trends you found in your investigations.

You will also include your understanding of the science readings and your discussions with each other. You may use the text in this book to help you write about the science you have learned. However, make sure to put it into your own words. The class will fill in the large *Project Board*. Make sure to record the same information on your own *Project Board* page.

How can you prevent good friends from getting sick?

What do we think we know?	What do we need to investigate?	What are we learning?	What is our evidence?	What does it mean for the challenge or question?

Project-Based Inquiry Science

Learning Set 2

What Kinds of Things Make You Sick?

The next question you might think about is *What kinds of things make you sick?* You know germs can make you sick. They cause many diseases. You may have already added the question, "What are germs?" to the *Project Board*. There are many types of germs. You may know about germs, but you may not know the different types of germs that exist. You are now going to learn about two groups of germs: bacteria and viruses.

You know what a communicable disease is and you know that colds are one type of communicable disease. You will learn more about specific communicable diseases and how they are spread. In this *Learning Set*, you will be introduced to different diseases. You will learn about the symptoms of each disease, who can get sick, how you can avoid getting sick, and how to cure the disease.

All this information will become part of a *Communicable-Disease Information Table* you will complete. As you fill in diseases on this chart, you will be on your way to answering the *Big Question*: *How can you prevent your good friends from getting sick?*

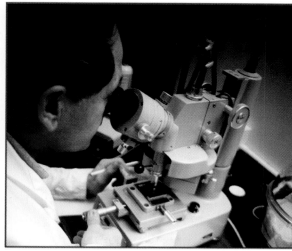

Germs are only visible under a microscope.

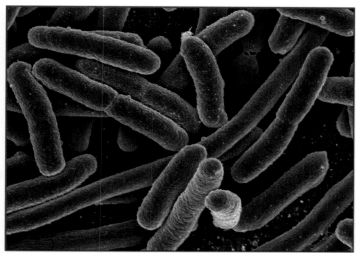

Escherichia coli *bacteria.*

2.1 Understand the Question

Thinking about What Kinds of Things Make You Sick

The question for this *Learning Set* is *What kinds of things make you sick?* It is a good idea to think about what you already know about what makes you sick. It is also important to think about what you are unsure about and what you would like to investigate.

Get Started

Think about the question. Share your ideas about what makes you sick. What do you already know about germs? What are the different types of germs? How are they different? Listen carefully to all the ideas presented. You may want to write down some of the ideas you hear. Think of several questions that might help you answer the question for this *Learning Set*. On your own, develop two questions that might help you understand how you get sick. When you write your questions, keep in mind that your questions should

- be interesting to you;
- require several resources to answer;
- relate to the *Big Question*; and
- require collecting and using data.

Your questions should not have a yes/no or a one-word answer.

When you have completed your two questions, take the questions back to your small group. Share all the questions with each other. Carefully consider each question and decide if it meets the criteria for a good question. With your group, refine the questions that do not meet the criteria. Choose the two most interesting questions to share with the class. Give your teacher the rest of the questions so they might be used later.

Update the *Project Board*

You will now share your questions with your class. Be prepared to support your questions with the above criteria. Your teacher will help you with the criteria if needed. Your teacher will then add your questions to the *Project Board*. In this *Learning Set,* you will work to answer some of these questions.

2.2 Read

Learn about Other Communicable Diseases

When you are sick, your body responds to the sickness. When you have a cold, you get a stuffy or runny nose and maybe a cough. The stuffy or runny nose and cough are called **symptoms** of the cold. All diseases have symptoms. Some diseases are easy to detect because their symptoms are very obvious. A cold is easy to detect. However, some diseases have symptoms that are less obvious or are just like the symptoms of other diseases.

Washing your hands can help to prevent the spread of disease.

Other parts of your body that get sick are your stomach and your intestines. **Gastrointestinal** illnesses are common illnesses that happen in your stomach or intestines. These illnesses have very obvious symptoms. The most common is **diarrhea**, but they can also make you vomit and give you abdominal (belly) cramps.

Both colds and gastrointestinal illnesses are communicable diseases. They are spread from one person to another. Think back to the glow-powder activity. Think of the powder as germs that can cause an illness. What made the "germs" move from one person to another? Germs that cause many illnesses can move from another person to you because they get on your hands. Then you may put your hands near your face. This is a way colds, gastrointestinal illnesses, and other sicknesses move from one person to another.

Read about diarrhea on the next page. As you are reading, pay attention to words you might not understand. Record these words as you are reading. You will find out what they mean as the Unit goes on.

symptom: an indication in your body that you have a disease.

gastrointestinal: related to the stomach and intestines.

diarrhea: frequent and watery bowel movements.

Gastrointestinal Infections

What Are Gastrointestinal Infections?

Diarrhea is usually caused by an **infection** in your body's intestines. Infections in the intestines are called gastrointestinal infections. Germs like **parasites**, viruses, or bacteria can all cause gastrointestinal infection. Diarrhea can also be caused by other illnesses, medicine you are taking, or changes in diet. Diarrhea is communicable only when it is caused by germs.

Which germs are responsible for diarrhea depends upon where a person lives. It also depends upon **sanitation** and **hygiene** standards. Countries that have poor sanitation have frequent outbreaks of diarrhea. In developed countries like the United States, outbreaks of diarrhea are most often caused by food poisoning. Sometimes food poisoning happens when bacteria in food that is not handled, stored, or cooked properly gets into a person's body.

What Is Diarrhea?

When you hear the word diarrhea, you probably think about discomfort and embarrassment. Diarrhea is no fun. But nearly everybody gets it once in a while. When you have diarrhea, you have frequent and watery bowel movements.

How Long Are Gastrointestinal Infections Contagious?

Infections that cause diarrhea are highly contagious. They can spread from person to person through dirty hands, contaminated food or water, and some pets. Most cases are contagious for as long as a person has diarrhea. However, some infections can be contagious for even longer.

Can You Prevent These Infections?

The most effective way to prevent the spread of contagious diarrheal infections is to wash your hands often. Dirty hands carry germs into the body. This can happen when you do things like bite your nails or use your hands when eating. It is important to always wash your hands well with soap and water after using the bathroom and

infection: the growth of germs in your body.

parasite: organism that lives and feeds either inside of or attached to another organism and does harm to that organism.

sanitation: the disposal of sewage and waste.

hygiene: things people do to stay healthy.

dehydration: (medical) a condition in which the body does not have enough fluid to function properly.

Drinking untreated water can cause food poisoning.

before eating. Washing your hands is especially important if you know there is a disease going around. Keeping bathroom surfaces clean can also help to prevent infections.

Food and water can also spread germs that cause diarrhea. To help protect yourself, cook foods thoroughly and wash raw fruits and vegetables well before eating them. Avoid eating undercooked hamburgers and poultry or raw eggs, and always refrigerate leftover food quickly. Make sure to clean your kitchen counters and cooking utensils, especially after they have been in contact with raw meat, eggs, and poultry.

Pets, particularly reptiles, can also spread germs. (Reptiles include animals like lizards, snakes, and turtles.) Pets should be kept away from family eating areas. Never wash pet cages or bowls in the same sink that your family uses to prepare meals. And always wash your hands after handling your pet!

How Are These Infections Treated?

Most infections that cause diarrhea will go away without treatment. Resting at home and drinking plenty of fluids to avoid **dehydration** are the best ways to get over the illness. Dehydration means your body does not have all the fluids (water) it needs. Sometimes doctors tell people to treat the symptoms of diarrhea by taking medicines to make their **stools** less watery. Other times diarrhea needs to be treated with **antibiotics**, or drugs that kill some germs.

Pets such as this bearded dragon can spread germs.

Stop and Think

1. The effects of an illness in your body are called symptoms. One symptom of a cold is a runny nose. What are the symptoms of gastrointestinal infections?

2. How do gastrointestinal infections spread?

3. How can you keep from getting a gastrointestinal infection?

4. You have learned a little about one gastrointestinal infection. Based on the reading, how can you make a good friend sick with a gastrointestinal infection?

5. What is one way you can prevent your good friend from getting sick with a gastrointestinal infection?

stool: the solid waste that is produced by the body.

antibiotic: a type of medicine that kills germs or prevents them from growing

Record Your Ideas

Throughout this Unit, you will record what you are learning about communicable diseases by entering the information in a table like the one shown. On your copy of the table, record what you now know about gastrointestinal diseases. Complete as many columns as you can. Some answers you may not know yet. You will be able to come back to the table later.

| | Communicable Disease Information Table | | | | | 2.2 |

Name: _____ Date: _____

Name of disease	Cause (bacterial or viral)	How does it spread?	Which body systems does it affect?	What are the symptoms?	How is it treated? Is there a vaccine?	Other notes
Write the scientific name and the common name. People usually say that they have "the flu," which is a shorter way to say that they have "influenza."	Many diseases are caused by either a virus or a bacterium.	Some diseases are spread through blood, body fluids (like saliva), air, contaminated food, and animal bites.	Some diseases affect several systems. List the main one(s).	The symptoms are what help you to determine that you have a particular disease.	"Treat the symptoms" means doing something so the symptoms will not be as bad. Sometimes, you can treat a disease, sometimes you can only treat the symptoms.	Use this column to write anything you think is interesting or important to remember about a particular disease.

PBIS_GF_SE_LS2_2-1BLM.doc

What's the Point?

You read about gastrointestinal diseases. Gastrointestinal diseases that are caused by germs are communicable. This means they can be passed from one person to another person. The reading described some of the symptoms of this disease. One of the symptoms is diarrhea.

You will enter information about each disease you learn about in a *Communicable-Disease Information Table*. The information will help you prepare to answer the *Big Question: How can you prevent your good friends from getting sick?*

2.3 Explore

Observing Cells

What structure do all living things, whether they are germs or humans, have in common? The **cell** is the basic unit common to all living things. Every living thing is made of cells. Most germs are made of a single cell. You are made of about 100 trillion cells. To understand how germs spread disease and how they affect the cells in your body, you need to know the structure and the functions of a cell.

Cells are the "building blocks" of all plants and animals. They are very small. Most cells are too small to see with the human eye. To actually observe them, you have to make them appear larger. To observe cells, you will look at them through a microscope. A microscope **magnifies** objects. It makes things appear much larger than they are.

Before you begin your observations, read about what a cell is and how it functions. Also, read about the parts of a microscope and their function.

cell: the structural and functional unit of all living organisms. (What a living thing is made of and what makes it work.) It is sometimes called the "building block" of life.

magnifies: makes something look larger, but does not actually enlarge the physical size of the object.

cell theory: a theory about the relationship between cells and living things.

The Cell Theory

After scientists were able to study cells for some time, they put together a theory, called the **cell theory**, about cells and living things. The cell theory explains the relationship between cells and all living things, large and small. The theory states that

- all living things are composed of cells.

- cells are the basic units of structure and function in living things.

- all cells are produced from other cells.

The cell theory helps scientists learn more about living things by studying cells. Because all cells come from other cells, scientists can study cells to learn about reproduction and growth.

Human blood cells under a microscope.

GOOD FRIENDS and GERMS

PBIS

organelle: a specialized structure in a cell.

nucleus: the control center of the cell.

chromosomes: contain the genetic material of the cell.

genetic material: contains the information that determines the traits of an organism; hereditary material.

cell membrane: surrounds the cell; controls the movement of materials into and out of the cell.

Cells – The Basic Units of Life

Scientists have spent many years observing cells. They have found that there are many different sizes and shapes of cells. Some different types of cells are shown in the photographs. Although cells come in different sizes and shapes, they have some structures in common.

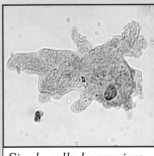

Single-celled organism Amoeba proteus

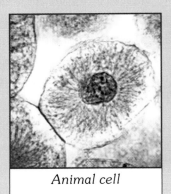

Animal cell

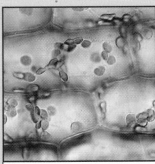

Plant cells

Each different structure in a cell has a specific function (role). Each structure acts as if it were a special organ. That is why cell structures are known as **organelles**.

The diagram shows some of the organelles that are found in animal cells. All these structures are also found in plant cells.

Animal Cell

cell membrane

chromosomes

nucleus

cytoplasm

mitochondria

vacuole

The **nucleus** is the control center of the cell. It holds the information that directs what the cell does. The nucleus contains **chromosomes**. Chromosomes store the **genetic material**. Genetic material determines the specific traits of an organism. The **cell membrane** surrounds the cell. It controls the movement of materials into and out of the cell.

The **cytoplasm** is the watery fluid inside the cell. It contains all the cell's organelles. **Mitochondria** (singular, mitochondrion) provide the cell with energy. In the mitochondria, food and oxygen are converted to energy, carbon dioxide, and water by the process of **cell respiration**. The **vacuole** is a storage area. It is used to store water and nutrients.

Plant Cells

Plant cells have all the same organelles found in animal cells. They also contain a few structures not found in animal cells.

In plants, the organelles in which food is made are called **chloroplasts**. The chloroplasts contain a green pigment called **chlorophyll**. Chlorophyll is used in **photosynthesis**. You may recall from previous science classes that photosynthesis is the process by which plants use light energy, water, and carbon dioxide to make sugar and oxygen. Plant cells also have a **cell wall**. The cell wall supports and protects the plant cell.

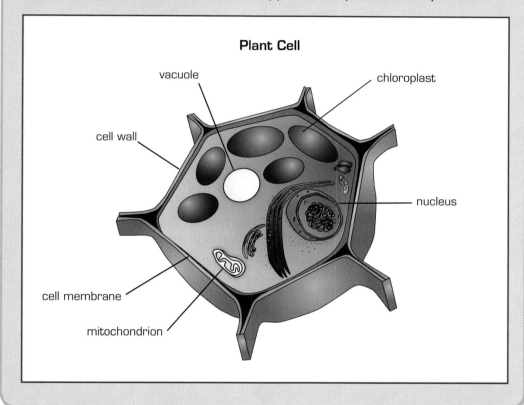

Plant Cell

vacuole

chloroplast

cell wall

nucleus

cell membrane

mitochondrion

cytoplasm: the watery fluid that contains the organelles of the cell.

mitochondrion (plural, mitochondria): provides the cell with energy.

cell respiration: the process by which food and oxygen are converted to energy, carbon dioxide, and water. vacuole: a storage area for food and water.

chloroplast: organelle that contains the green pigment chlorophyll used in photosynthesis.

chlorophyll: a green pigment.

photosynthesis: the process by which plants make sugar and oxygen using light, water, and carbon dioxide.

Stop and Think

1. Name four structures found in both animal and plant cells. What is the function of each?

2. What structures are found in plant cells that are not found in animal cells?

3. What can a plant cell do that an animal cell cannot? What structure carries out this function?

cell wall: protects and supports the plant cell.

compound light microscope: a microscope that has more than one lens and uses light transmitted to your eye to form an image.

The Microscope and Magnification

You cannot see most cells with the unaided eye. You need a microscope to observe them. There are different types of microscopes. The type you will be using is called a **compound light microscope**. The diagram shows the parts of this type of microscope.

The *stage* holds the microscope slide. The *clips* hold the slide in place. An opening in the stage lets light shine through. The *ocular lens* magnifies the object. The *objective lenses* found on a *revolving nosepiece* also magnify the object. There are usually three objective lenses. The shortest, or low-power lens, usually magnifies the object four times (written as 4×). The medium-power lens usually magnifies the object 10×. The longest lens, the high-power lens, usually magnifies the object 40×. The *coarse-adjustment knob* moves the *body tube* up and down. It is used only with the low-power lens. The *fine-adjustment knob* also moves the body tube up and down. It is used with the medium- and high-power lenses.

You have probably used a handheld magnifying glass to make objects appear larger. A magnifying glass, also called a hand lens, is a single lens that magnifies objects. An important development of the microscope was to use two lenses. Each lens multiplies the magnification of the other. If a lens magnifies an object 10×, the object appears 10 times larger. If you then add another lens that also magnifies the object 10×, the object now appears 100 times larger (10 × 10)!

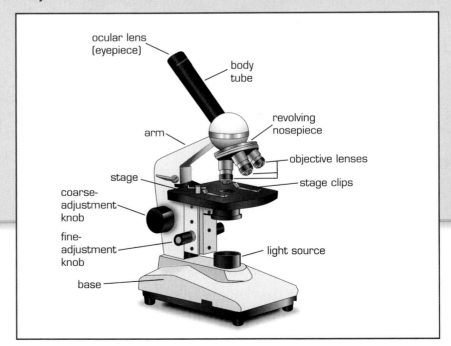

Procedure: Observing Animal Cells

You will be observing animal cells under the microscope. Scientists work hard to make accurate observations. Careful observations help scientists make predictions, draw conclusions, and find comparisons. It is very important that you take time to make careful observations when looking at the cells. Do not rush through your observations. Another important scientific skill you will need to use in making your observations is sketching.

1. Make sure that the nosepiece of the microscope is set so the low-power lens is in place. Put the slide on the stage. Hold it in place with the clips.

2. Look at the microscope from the side. Find the coarse-adjustment knob. Use it to lower the low-power lens. Bring it as close to the slide as possible. Do not let the lens touch the slide.

3. Look at the slide through the eyepiece. Very slowly move the coarse-adjustment knob away from the slide. Turn the knob until the object on the slide comes into focus (becomes clearer).

4. On the *Observing Cells* page, make a sketch of what you see as you look through the low-power lens. You should try to sketch exactly what you see. Make the object fill the same amount of space in your sketch as it does in the microscope. Be careful to include as many details as you can see. Suppose you see a big circle in the middle of the cell. In your diagram, sketch the big round circle and put it in the middle of the cell. If there are many smaller things in the middle of the cell, you should sketch them all. Try sketching them in the same places where you see them.

5. Look through the microscope again. Adjust the slide so the object you are looking at is in the center. Rotate the nosepiece to the medium-power lens. Use the fine-adjustment knob this time. Bring the slide into focus.

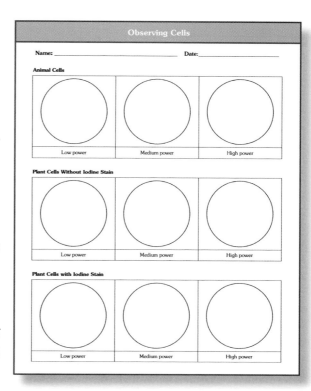

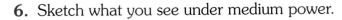

6. Sketch what you see under medium power.

7. Repeat Steps 5 and 6 using the high-power lens.

Procedure: Observing Plant Cells

You will now look at a type of plant cell, an onion cell. You will first prepare a wet-mount slide of the onion skin cell and then look at it under the microscope. Your teacher may also give you prepared slides of other plant cells to observe.

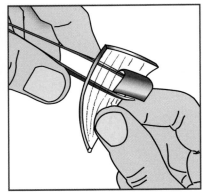

1. Get a piece of onion skin. Using tweezers, remove a thin, transparent layer from the inside surface. Transparent means that you should be able to see through the layer that you remove.

2. Place the piece flat in the center of the slide. Be careful that the piece does not fold over.

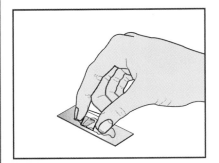

3. Cover the onion skin with one or two drops of water. Gently lower a cover slip over the slide, as shown in the diagram.

4. Observe the slide under the microscope. Sketch what you see under each magnification on the *Observing Cells* page.

5. Switch to low power and remove the slide.

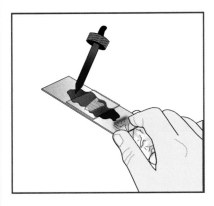

6. Put on rubber gloves and safety goggles. Place a drop of iodine stain at one edge of the cover slip. Hold a piece of paper towel at the other end of the cover slip. The paper towel will draw the iodine stain under the cover slip and across the onion skin.

7. Observe the slide under the microscope again. Sketch what you see under each magnification on the *Observing Cells* page.

8. You may observe other prepared slides of plant cells under the microscope.

Stop and Think

1. Describe in words the animal cells you looked at.

2. Describe in words the plant cells you looked at:
   a) without the iodine stain.

   b) with the iodine stain.

3. Why do you think you used the iodine stain? What effect did it have on the cells?

4. Which cell parts were easy to see and which cells parts were difficult or impossible to see in
   a) the animal cell?
   b) the plant cell?

5. Which cell parts looked similar in all the cells you observed?

6. Which cell parts looked different in all the cells you observed?

7. What parts did you observe in both animal and plant cells?

8. What differences between the animal and the plant cells did you observe?

Conference

In your small group, share all the sketches you made. Identify the parts of the cells and label them with the proper names. Check with the members of your group to see if you have labeled the cells the same way.

Create a poster for a presentation. Choose one cell picture from those in your group and draw it on your poster. Make the picture as similar to the original sketch as possible. Label the organelles that you could see with the microscope. Write the job or function of each organelle next to the label on the poster.

Be prepared to present your poster to the class.

Communicate Your Results

Investigation Expo

Place your poster on the wall. As you observe the other posters, notice any differences in the ways your classmates sketched their cells. After everyone has had a chance to see all the posters, discuss any differences you noticed.

Update the *Project Board*

You have just observed animal and plant cells. You know that it is important to learn about cells in order to answer the question: *What kinds of things make you sick?* Focus on the two columns on the *Project Board: What are we learning?* and *What is our evidence?* Record what you have learned about cells. Make sure to record the same information on your own *Project Board* page.

What's the Point?

It is important to know about the cells in your body. Cells move disease around and cells keep you healthy. All the different cells in your body have specific jobs, called functions.

Most cells are too small to see with only your eyes. You need a microscope. The parts inside cells are even smaller. The parts, called organelles, each have a function. They work together to keep the cell functioning properly.

It is important to know about the structure and functions of cells to understand what makes you sick and how you can keep from getting sick. Most germs are cells. All your body parts are made of cells. You learned that there are parts that are common to all cells. That means that cells that make you sick and cells that keep you healthy are very similar to each other.

Technology Connection

Science and technology work together. Until a microscope was made, no one could see a cell. As better microscopes were developed, scientists were able to see more. They learned more about the parts of cells and their functions. This led the scientists who made microscopes to make even more powerful ones.

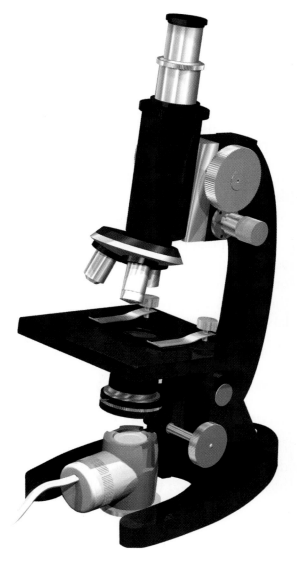

More to Learn

How Big Is Very Small?

Cells are very small. You cannot see most cells without using a microscope. You looked at plant and animal cells under a microscope. Exactly how small were the cells you looked at? Can you use a microscope to estimate the size of the cells?

If you can calculate the size of the microscope's **field of view** using a ruler and estimate the number of cells that you can fit across the field of view, you can estimate the size of each cell. The field of view is the circle of light you see through the microscope.

The first procedure below guides you through the steps to find the **diameter** of the field of view of your microscope under low and medium power using a ruler. Most high-power lenses have a field of view that is less than one millimeter, the smallest marking on your ruler. You will not be able to use a ruler for the high-power lens. You will use a **ratio**. You will then estimate how many objects could fit into each field to determine the size of the object.

Procedure: Determining the Field of View

1. With the low-power lens in place, put a transparent ruler on the microscope's stage. Position the millimeter marks on the ruler below the lens. Center the ruler in the field of view.

2. Using the coarse-adjustment knob, focus on the marks on the ruler.

3. Move the ruler so that one of the millimeter markings is just at the edge of the field of view. Record the diameter of the field of view in millimeters. This is the size of the field of view of the low-power lens.

4. Use the same procedure to measure the field of view of the medium-power lens.

field of view: the circle of light you see when you look through a microscope.

diameter: the length of the line through the middle of a circle from one side to the other.

ratio: a comparison of two numbers.

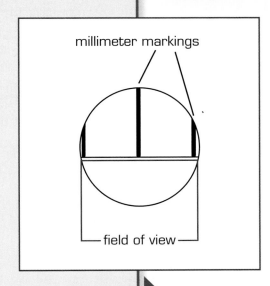

millimeter markings

field of view

5. The high-power objective lens on your microscope probably has a field of view that is less than one millimeter. You cannot measure this field of view using a ruler. You need to do mathematical calculations.

First, calculate the ratio of the magnification of the high-power objective lens to the magnification of the low-power objective lens.

$$ratio = \frac{magnification\ of\ high\text{-}power\ lens}{magnification\ of\ low\text{-}power\ lens}$$

Use this ratio to determine the diameter of the field of view under high-power magnification.

$$field\ diameter\ (high\ power) = \frac{field\ diameter\ (low\ power)}{ratio}$$

Procedure: Estimating Size

1. You now know the diameter of the field of view under the low-, medium-, and high-power lenses. You can use this information to estimate the size of an object you see under the microscope.

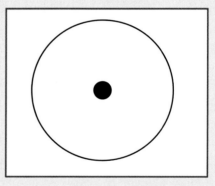

2. Look at the object under the low-power lens. Estimate how many of these objects would fit across the diameter of the field of view.

3. Calculate the width of the object.

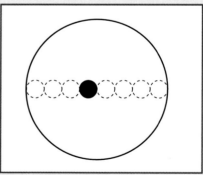

$$width\ of\ object = \frac{width\ of\ field\ of\ view}{number\ of\ objects\ across\ the\ diameter\ of\ the\ field\ of\ view}$$

2.4 Read

Some Germs Are Single-Celled Organisms Called Bacteria

The question you are investigating is *What kinds of things make you sick?* You know germs can make you sick. Some germs are **bacteria**. In this section, you will read about bacteria.

One-Celled Organisms: Bacteria

In the last section, you looked at several plant and animal cells. All the cells you looked at had a nucleus. Scientists call an organism made of cells that have nuclei a **eukaryote**. A eukaryote can be made of one cell or many cells.

There is another type of organism called a **prokaryote**. All prokaryotes are made of a single cell. However, these cells do not have nuclei. Bacteria (singular, bacterium) are examples of prokaryotes.

Look at the picture. It looks like the cells that you were observing through the microscope. However, notice the difference. There is no nucleus.

Bacteria and other very small organisms are also called **microorganisms** or **microbes**. These are organisms that can only be seen through the microscope.

Bacteria are found everywhere, and they can survive almost anything. Some kinds of bacteria live in your stomach, which has a lot of acid. Bacteria live on kitchen counters that are cleaned often. Bacteria can live in the harshest places in the world. They have been found in polar icecaps. They also live in hot-water vents at the bottom of the ocean.

There are many different kinds of bacteria. There are over 300 kinds of bacteria in your mouth alone. In your body, there are 10 times more bacteria than there are cells in your body.

bacteria (singular, bacterium): the most common form of one-celled organisms.

eukaryote: an organism whose cells have nuclei.

prokaryote: a single-celled organism that does not have a nucleus.

microbe (or microorganism): an organism that can be seen only through a microscope.

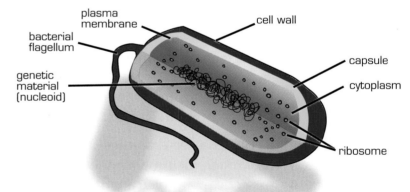

plasma membrane
bacterial flagellum
genetic material (nucleoid)
cell wall
capsule
cytoplasm
ribosome

Bacteria: Good or Bad Guys?

You know that some bacteria can make you sick. However, some bacteria in your body actually keep you healthy. Some bacteria in your digestive system keep you healthy by breaking down food materials. Other bacteria in your digestive system produce vitamins that your body needs to stay healthy.

People also use bacteria for a wide variety of purposes. The dairy industry uses bacteria when making cheese, yogurt, sour cream, and other products. Bacteria are also used to break down waste in sewage-treatment plants, clean up oil spills, and develop antibiotics.

Bacteria play important roles in the natural world. Bacteria live in the soil. They help plants grow successfully. Bacteria help make gardens and farms more fertile. They make it possible for elements such as carbon, nitrogen, and oxygen to be returned to the atmosphere by decomposing waste and dead organisms. If there were no bacteria, the world would not look the way it does.

Are Bacteria Alive?

Bacteria are single-celled organisms. Just like all other organisms, bacteria need energy and materials to stay alive, grow, develop, and reproduce. They obtain the energy and the nutrients they need from the food they eat. Most bacteria obtain their food from other living organisms. This is how many of the disease-causing bacteria obtain their food. You are one of the organisms from which they obtain food. Some bacteria obtain their food from dead organisms, and a few can make their own food.

Bacteria use **enzymes** to break down this food. They take up nutrients through their membranes. Bacteria then convert nutrients into energy. The process by which animals break down their food to obtain energy and then use the energy in their life processes is called **metabolism**.

Bacteria get larger by using energy to grow. Each cell grows larger. Then it divides into new cells. One bacterium will grow and divide into two bacteria. This is a way bacteria reproduce. Then each of the two bacteria divides, making four bacteria cells. Each of these four bacteria cells divides to make eight. These divisions can happen very quickly in a favorable environment. What a favorable environment is varies among the different bacteria. After many divisions, a colony of bacteria has developed. There are many individual bacteria cells in a colony. Bacteria continue to divide until they have used up

---

Did You Know?

Parasites are organisms that benefit by living on or in another organism. Some bacteria and **fungi** are parasites. Fungi are eukaryotes that have cell walls, reproduce by **spores**, and get food by absorbing it from their surroundings. One type of fungus causes athlete's foot, a disease that affects the outer layer of human skin.

fungi (singular, fungus): eukaryotes that have cell walls, reproduce by spores, and get food by absorbing it from their surroundings

enzyme: a substance that causes a chemical change in another substance.

all their food or created too much waste to survive.

Like all other organisms, bacteria need to adjust to a changing external environment. They need ways to keep their internal environment —the amount of fluid, the materials, and the temperature—inside the cell fairly constant. The process of keeping a constant internal environment is called **homeostasis**. Bacteria have developed various ways of doing this.

One important way that many bacteria respond to their environment is **spore** formation. When conditions for bacterial growth are not favorable, these bacteria form a tough capsule around their genetic material and some cytoplasm, forming a spore. The spore can remain **dormant** for years. When conditions return to favorable, the bacteria will begin to grow and reproduce again. Spore formation allows some bacteria to survive extremely harsh conditions—lack of food, extreme temperatures, dryness, and chemical **disinfectants**.

Look at the pictures of different types of bacteria. Most bacteria have one of three shapes: rod-shaped, spherical, or spiral. The string-like parts in the bacteria in the bottom picture are called **flagella** (singular, flagellum). The bacterium uses those to move.

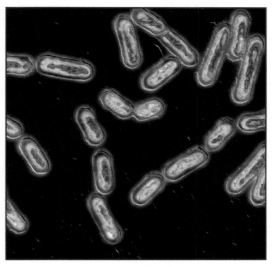
Rod-shaped bacteria (bacillus).

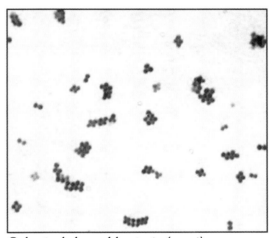
Spherical-shaped bacteria (cocci).

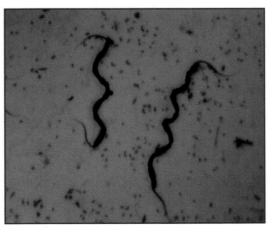
Spiral-shaped bacteria (spirilla).

metabolism: the combination of chemical reactions that takes place in an organism; food is converted into energy that the organism uses to carry out its life processes.

colony: a group of only one kind of bacteria that grows from a single, original bacterium.

homeostasis: the process by which an organism keeps its internal environment in a constant condition despite changes in its external environment.

spore: in bacteria, a dormant structure that allows the bacterial cell to survive unfavorable conditions. In fungi, a cell that develops into a new organism.

dormant: temporarily not active.

disinfectant: a substance that can kill microorganisms.

flagellum (plural, flagella): a string-like part on some cells that helps the cell move.

How Do Bacteria Make You Sick?

toxin: a poisonous substance.

toxic: poisonous.

protein: a substance that is found in all cells; it is necessary for life.

What happens after bacteria are in your body depends on the type of bacteria. Some cannot survive inside you, so they cannot harm you. But other bacteria can survive. The bacteria that survive begin to reproduce inside you. In larger numbers, they can affect your body. Some of these bacteria produce **toxins** (**toxic** substances). Toxins can have different effects on your body. Some can make you vomit or cause diarrhea. Others can affect nerve cells and prevent signals from passing through your nerves. This can cause paralysis. Other bacteria can actually invade and destroy your cells. One type of bacteria that attacks and destroys the cells of your intestines also causes severe diarrhea. Some bacteria can produce specific **proteins** to which your body reacts. Your body's reaction to the protein makes you sick. As you investigate the disease you will select, you will find out how the type of bacteria that causes the disease affects the human body.

Stop and Think

1. Why are some organisms called microorganisms or microbes?

2. What is the difference between a eukaryote and a prokaryote?

3. What type of organism is bacteria, a eukaryote or a prokaryote?

4. Describe three ways that bacteria benefit humans.

5. Why is it important to know that some bacteria can form spores when you are thinking about how to prevent others from getting sick?

6. Explain two ways that bacteria can make you sick.

7. Describe a situation in which a bacteria-free environment would be

   a) desirable.

   b) not desirable.

What's the Point?

Bacteria are single-celled organisms. Like all living things, bacteria need to maintain a constant internal environment. This process is called homeostasis. Bacteria have three basic shapes — rod, spherical, or spiral. Bacteria can be found in most places in the world. Without bacteria, the world would not work as it does. The bacteria that live in your body do important jobs to keep you healthy. Some bacteria can be very dangerous and can make people sick by releasing toxins or destroying tissues.

2.5 Case Study

Analyzing Catering-Service Sickness Data

In this section, you are going to read about Darrell. Darrell and his uncle are trying to find out why customers of his uncle's catering business are getting sick. Read the case carefully. As you are reading, be sure to use the *Communicable-Disease Information Table* as a resource to understand what could be making the people who buy food in Darrell's uncle's restaurant sick.

Examine a Case Study

One summer, Darrell got a job in his Uncle Joe's restaurant. His uncle makes great chicken and coleslaw. Near the restaurant is a city park. Each summer, families meet at the park for reunions. They often purchase chicken and coleslaw from Darrell's uncle for their reunions. Darrell's uncle was concerned because several people had called the restaurant the previous summer to complain about getting sick after eating his restaurant's food.

Darrell's uncle uses the freshest products to make his chicken and coleslaw. He is not sure why people are getting sick. Darrell wonders, "What is causing people to get sick at the family reunions?"

Darrell knows that bacteria on food can cause gastrointestinal illness. He knows that bacteria reproduce quickly in a warm environment. Darrell's uncle asks people to keep their coleslaw and chicken in a cooler during the reunion. The cooler helps keep the food at the right temperature for safety. But maybe all the people do not follow his instructions. Darrell decides to collect data on how the food is stored and at what temperature it is kept. He thinks it is also important to know the outside temperature.

Darrell has noticed that sometimes people put the food out on the table rather than leaving it in the cooler. Every time he makes a delivery to a reunion, Darrell watches to see if the food is stored in a cooler or placed on the table. He records the daily outside temperature and where the food is stored. In addition, if people call to complain, he politely asks them how many people got sick at the party. After collecting data for three weeks, he shared it with his uncle. Darrell thought a table would be a good way to display his information.

Uncle Joe's Catering Service Sickness Data

Date	Observations about where food is stored	Outside temperature	Number of sick people
July 1	In cooler	77°	0
July 3	On table	76°	1
July 5	On table	80°	2
July 7	On table	82°	4
July 9	In cooler	82°	1
July 11	In cooler	83°	0
July 13	On table	81°	1
July 15	On table	86°	4
July 17	On table	84°	2
July 19	In cooler	84°	0
July 21	On table	87°	6

Stop and Think

1. What variables did Darrell track for his investigation?

2. Why did Darrell record the outside temperature?

Analyze Your Data

The table shows the data Darrell collected. The data should be organized so Darrell's uncle can make sense of it to answer the question. He needs to be able to see differences or make comparisons between different variables. Organize the data so that you can make comparisons or see differences.

Then make three statements about what the data tells you about why the people were getting sick. What will you tell Darrell's uncle about the food to keep other people from getting sick? Record your recommendation and be prepared to share it with others.

Reflect

1. Why do you think your way of organizing information would help Darrell's uncle understand the data so he could answer the question?

2. What does this investigation have to do with bacteria and spreading disease?

3. Can you tell from the case study if the illness people got might be a communicable disease?

Communicate Your Results

Investigation Expo

In your small group, discuss what conclusions you can draw from the data. Record your conclusions and be ready to present them to the class in an *Investigation Expo.*

As a class, review and discuss the conclusions drawn by each group. What data did they use to support their conclusions? Come up with conclusions the whole class agrees upon.

GOOD FRIENDS and GERMS

Conference

The *Project Board* is beginning to take shape. You need to add the information you found about bacteria and cells to the *Project Board*. What information is most important to add to the board? Write three things you think are important. Next to each of your ideas, write in which column of the *Project Board* you think your information belongs.

Also, now that you have learned more about cells and bacteria, write two new questions. What questions are you interested in answering?

Update the *Project Board*

Share the three things you think are important to add to the *Project Board* with the class. As a class, decide which information about cells and bacteria is important to add to the *Project Board*. If you add a fact to the *What are we learning?* column, be sure you have evidence to support it. Record this in the *What is our evidence?* column.

Then discuss the questions that you and your class think need to be answered about what makes you sick. Record these in the *What do we need to investigate?* column. Keep a record of what was added to the *Project Board* on your own *Project Board* page.

What's the Point?

In the last section, you learned that under favorable conditions, bacteria reproduce and multiply very quickly. Bacteria that cause disease multiply quickly in warm, moist conditions. The food that Darrell's uncle sold probably contained very few bacteria. However, if the customers did not keep the food cold, a few bacteria would very quickly have grown to become many bacteria in warm temperatures. When the customers ate the food containing all these bacteria, they got sick.

2.6 Investigate

Where in Your Classroom Can You Find Bacteria?

You now know that harmful bacteria can be anywhere. There may be bacteria in your classroom. In this section, you will work as a class to develop an investigation that shows the presence of bacteria in your classroom. Then in the next section, you will follow a similar procedure to do an independent investigation to answer a question that interests you.

Microbiology is the study of microorganisms. Over many years, **microbiologists**, people who specialize in microbiology, have developed very specific techniques and procedures to study microbes. Before your class designs their investigation, read the sample procedure for growing bacteria. The procedure uses microbiology techniques adapted for a classroom.

microbiology: the study of microorganisms.

microbiologist: a scientist who specializes in microbiology.

Materials
- **Petri dish with nutrient agar**
- **cotton swabs**
- **permanent or wax-pencil marker**
- **adhesive tape**

Sample Procedure for Growing Bacteria

Scientists and technologists who study bacteria have certain materials and procedures they use. The materials listed here are similar to those they use. The list has materials that are suitable for your classroom. Here is a brief outline of a procedure that could be used for growing bacteria.

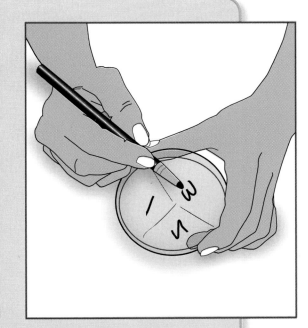

1. Use a wax pencil or permanent marker to divide the bottom half of a Petri dish containing nutrient agar into three sections. (Nutrient agar is a jelly-like substance that contains the nutrients bacteria need to grow.) By dividing the dish, you can use it for samples from more than one area. Carefully label each section so you know which area your bacteria were collected from.

GOOD FRIENDS and GERMS

Do not open the Petri dishes after you have swabbed them.

You do not know what microbes are growing on the agar. They may be dangerous to your health.

All Petri dishes must remain tightly sealed. Your teacher will dispose of all the dishes when you have completed your investigation.

2. Use a cotton swab to collect a sample of bacteria from an area of interest. Wipe the area you want to test with the swab. Use a different cotton swab for each area you test. Be very careful that the end of the cotton swab does not touch you or anything else you are not testing.

3. Carefully lift the lid from the Petri dish. Lift up one side only enough to be able to gently swab the surface of the agar plate. Try not to breathe onto the surface.

4. Gently brush or touch the cotton swab to the surface of one of the labeled areas of the agar plate.

5. Repeat Steps 2 to 4 for the other areas you have decided to test.

6. Securely seal the lid to the Petri dish using adhesive tape.

7. Invert the Petri dish and place it in a warm place.

8. Check the Petri dish every 24 h (hours). Record your data.

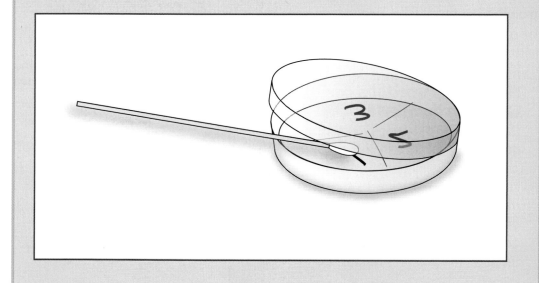

Plan Your Investigation

Read the questions in each part of the *My Experiment Page*. The purpose of the questions is to help you think about how to design an experiment. You used some of the questions on the *My Experiment Page* to guide previous investigations. You will use these questions to plan and guide the investigation for this section.

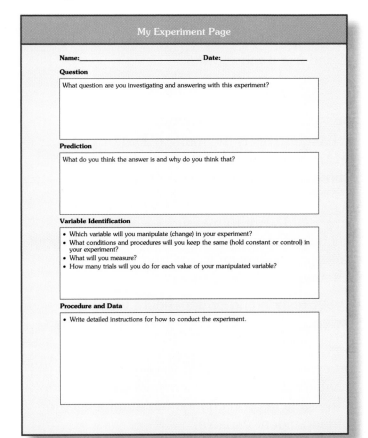

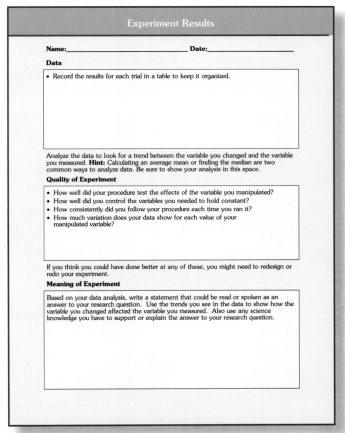

Question

- What question are you investigating and answering with this experiment?

With your class, talk about the different places you think there are a lot of bacteria. Then think of a question you could investigate about that place and bacteria. Record the question your class develops.

For example, some possible questions might involve classroom desks, pencils or pens, hands, the water fountains, or doorknobs. Your question might be: *Does my right hand or my left hand have more bacteria on it?*

Prediction

- What do you think the answer is, and why do you think that?

You need to make a prediction and provide some reasons for your prediction. Look again at the question. What do you think the answer to the question is going to be? Record your answer, and then describe why you think that is going to be the answer. Be prepared to share and discuss your prediction with the class.

For example, one prediction might be "My left hand will have more bacteria on it because I am left-handed and use it more often."

Variable Identification

You will need to identify all the variables in your experiment.

- What variable will you change? This is the manipulated or independent variable. You should only change the value of one variable. It may be the area from where you collect bacteria.

- What variables will you measure? These are the responding or dependent variables. You might want to measure or observe the things that you think will be different depending on the value of your independent variable. You probably want to measure how much bacteria grew from each area you decided to test.

- How will you measure your responding or dependent variables? You need to decide how you will measure the amount of bacteria in each location.

- What conditions and procedures will you need to control in your experiment? That is, what conditions and procedures will you keep the same as you change the manipulated or independent variable?

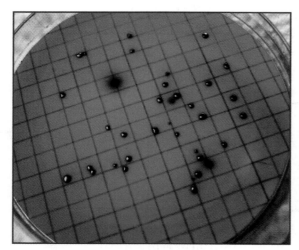

Bacteria growing in nutrient agar.

To answer the question, "Does my left or right hand have more bacteria?" your manipulated or independent variable will be the hand you collect from. The responding or dependent variable will be the amount of bacterial growth on the Petri dish. You will measure the responding or dependent variable by measuring the diameter of the bacterial growth. You will control the place on each hand from which you collect the bacteria.

Procedure and Data

Using the sample procedure as a guide, with your class, identify the steps of the investigation. Record detailed instructions for how to conduct the experiment. Someone else should be able to follow your instructions and duplicate your experiment. Include how to set up your experiment, how you will gather samples and record your data, how to measure the changes in your samples, and how to analyze your data.

Be careful when you are writing your procedure. Make sure you write it very accurately. Clearly record each step of the procedure. Make it clear enough that someone else could follow it. Pay attention to each part of the procedure and data list.

Run Your Experiment

Carry out your procedure. Be sure to follow the steps carefully. Record your results.

Analyze Your Data

Answer the following questions. As a class, discuss your answers.

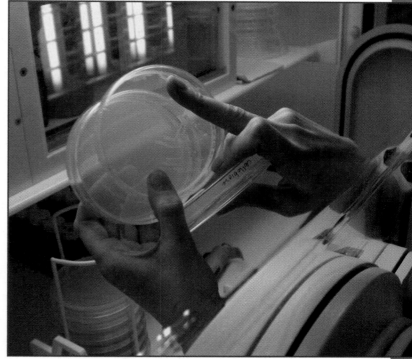

Scientists grow bacteria on agar in a Petri dish.

1. Which area(s) that you tested had the most bacteria? What evidence do you have to support your claim?

2. Which area(s) that you tested had the least bacteria? What evidence do you have to support your claim?

3. How did the results of your investigation compare with your prediction? Suggest reasons for any differences. If your results surprised you, tell why.

4. Make a claim about the bacteria on the surfaces you tested. Your claim should answer the research question your class developed. Support your claim with evidence from your investigation.

Reflect

Think about the following questions. Discuss your answers with the class.

*   How reliable are your results? Support your answer with evidence.

*   What are some of the sources of error that you may have introduced? For example, did all the colonies of bacteria that grew on the agar come from the area tested?

*   In the next section, you will be planning and running your own investigation. What have you learned in this section that will help you design an investigation? What will you need to pay particular attention to in designing your investigation?

What's the Point?

In this investigation, you used microbiology techniques to answer your class's question about bacteria in your classroom. Microbiologists have spent many years developing procedures that produce reliable results. When they test an area for bacteria, they are very careful not to introduce bacteria from other sources onto the surface of the agar.

The investigation helped you think about where bacteria might be in your environment. The *Big Question* for this Unit is *How can you prevent your good friends from getting sick?* You probably found that there were bacteria in many parts of the classroom. Some of those bacteria could make you and your friends sick.

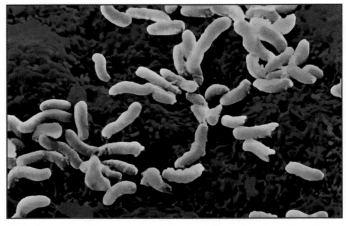

Heliobacter pyloris *are rod-shaped bacteria (artificially colored) that cause stomach ulcers in people.*

Your observations are similar to the observations that scientists make when they do investigations. Epidemiologists do investigations to track germs. Then they use the evidence from their investigations to make recommendations that could prevent people from getting sick.

2.7 Investigate

Where Else Do Bacteria Live?

The investigation you just finished provided you with some evidence about where bacteria grow in your classroom. You may have been surprised by the results. What questions would you like to ask about bacteria based on what you have observed? You will now use what you observed about bacteria and designing an investigation to plan and run an investigation to answer your own research question.

Plan Your Experiment

There may have been something about the data from the last investigation that surprised you. You can turn this into a question that you can answer in this investigation. For example, you may have thought that a metal doorknob would have a lot of bacteria on it because so many people touch it. Instead, perhaps you found that there were less bacteria on the doorknob than on your desk. You could turn this into a question: "Are there more bacteria on a wooden surface than on a metal surface?"

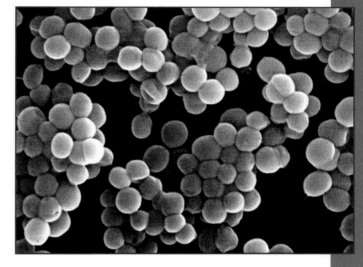

An area of skin as small as 6.5 cm² (1 square inch) may be home to more than half a million bacteria, such as Staphyloccus epidermis *that is shown here.*

Also, in Section 2.4, you generated new questions about bacteria. You can use some of these questions to help you develop a research question for this investigation.

With your group, discuss what research question you would like to investigate. Working within your group, develop a question, and record it on your *My Experiment* page.

Plan an investigation to answer your group's question. Use the *My Experiment Page* to help you plan your investigation. You may wish to review the part in Section 2.6 that explains how to use the questions and criteria on the *My Experiment Page*.

Create a numbered list of steps that you will use to gather bacteria, grow bacteria, and judge the amount of bacteria that has grown. Think about the precautions you must take and why they are important to follow. After you have recorded your procedure, you will present it to the class.

Communicate Your Plan

Plan Briefing

Share your investigation plan with the class. Also, listen to the plans of other groups carefully. Think about the question they are going to answer and how their plan is going to work. Have they met the criteria for good experiment design? If not, what could they do to improve their plan?

Revise Your Plan

With your group, revise your experiment plan based on the discussion you just had in class.

Run Your Experiment

With the approval of your teacher, carry out your investigation.

⚠️

Your teacher must approve your procedure before you begin to run your experiment.

Analyze Your Data

After you have made your observations, discuss the results of your investigation with other members of your group. The questions on the next page should help you organize your discussion. Take notes on what others are saying during the discussion.

- What variable were you investigating? What were you investigating about that variable? How did you vary it to determine its effects?

- List all of the variables you tried to hold constant in your experiment.

- Do you think that the data set you collected was useful in determining the effect your variable had on the growth of bacteria? Describe why or why not.

- What is the answer to your research question? How confident are you of your answer?

Communicate Your Results

Investigation Expo

Recall that an *Investigation Expo* is designed to help you present results of an investigation. Scientists present results of investigations to other scientists. This lets the other scientists build on what was learned. Each group in your class investigated a different question. Others in your class will want to know your results to help them answer the *Big Question: How can you prevent your good friends from getting sick?*

There are several things your classmates will want to know about your investigation. These include the following:

- the question you were trying to answer in your investigation

- your prediction

- the procedure and what makes it a fair test

- your results and how confident you are about them

- your interpretation of the results and how confident you are of them

To prepare for an *Investigation Expo*, make a poster that includes all of the five items listed above. Present them in a way that will make it easy for someone to make sense of your poster. Others should be able to identify what you have done and what you found out.

There are two parts to an *Investigation Expo*: presentations and discussions. As you look at posters and listen to other groups present their work, ask questions you need answered to understand the results and to satisfy yourself that the results and conclusions others have drawn are trustworthy. Be sure that you trust the results that other groups report.

Reflect

Answer the following questions. Be prepared to share your answers with the class.

1. What did you learn about where bacteria live?

2. Why do you think they live in those places?

3. What trends did you see in the data of other groups about where bacteria live?

4. Think about the investigation you planned. Was it a fair test? If you do not think you ran a fair test as you had planned, describe how you would change your procedure if you had a chance to run the experiment again.

What's the Point?

The investigations in this section and the previous section helped you understand more about bacteria. You have learned about where they might be present. The data and conclusions you had for your investigations may have been difficult to make sense of. This is true in many science investigations.

Sometimes it is helpful to do an investigation more than once so that there is more data to analyze. Sometimes it is helpful to share your data with others so they can offer their ideas. This is why you do *Investigation Expos*. In these presentations, you hear about other people's results and combine them with your own to better understand the claims. This gives you the opportunity to see how many parts of the puzzle come together.

2.8 Read

Some Germs Are Viruses

Any organism or particle that can get inside of you and make you sick is an **infectious agent**. You learned that some, but not all, bacteria are infectious agents. A virus is another example of an infectious agent. Like bacteria, they are microbes. They cannot be seen by the unaided human eye. However, in many ways viruses are different from bacteria. Viruses are much smaller than bacteria. They cause different illnesses. They work differently. The illnesses caused by viruses are treated differently. Viruses, although very small, cause a lot of changes in your body. They can also affect other animals and plants.

What Are Viruses?

Viruses are particles that can cause many different types of diseases in plants and animals. Unlike bacteria, viruses cannot survive on their own. They need a **host** to survive. Viruses can only reproduce inside cells. Each virus has a way to reproduce once it is inside an organism's cell.

Viruses cause many different diseases. Some, like the cold virus, make you uncomfortable. Others, like the polio virus, can stop your body from working as it should.

There are many different types of viruses. Some viruses are able to infect only certain types of plants and animals. Some viruses infect only plants, others infect only animals, and still others infect only bacteria. For example, the bacteriophage infects bacteria. The feline immunodeficiency virus infects only cats.

infectious agent: something that can get inside your body, multiply, and cause disease.

host: an organism (animal or plant) that harbors (provides food and a place to stay) for another organism, such as a virus.

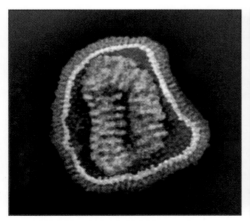
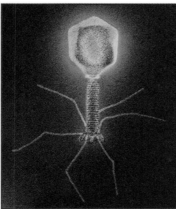

Viruses have many different shapes.

droplet transmission: a way that an infectious disease can be transmitted. Droplets containing an infectious agent (bacteria or viruses) are released into the air when an infected person sneezes, coughs, talks, or exhales. They then come into contact with another person's eye, nose, or mouth.

Because of the number of viruses and the ways they can change over time, it is difficult to treat illnesses caused by viruses (viral illnesses).

Treatment for illnesses caused by viruses depends on the illness. Some diseases, like the common cold and the flu, are hard to treat and impossible to cure. You might take some "cold medicine." The medicine will make you *feel* better. It may relieve your symptoms. However, it does not make you better. For that, you have to wait for the virus to die out. For a cold, this usually takes about seven days. A lot of illnesses caused by viruses are like that. Viruses are very hard to remove from your body. For most viral illnesses, you just have to wait until the virus dies out. The table gives you information about some common viral illnesses.

Viruses are easily moved from one person to another the same way bacteria are spread. Many viruses infect other people by **droplet transmission**. Other viruses are spread through contact with open sores.

Disease	Caused by and spread by	Effect on body	Cure or not?	Picture
influenza (also called "the flu")	caused by: influenza virus spread by: person to person in droplets of coughs and sneezes, droplet spread	discomfort, high fever, can lead to other diseases	no cure, treat the symptoms to make more comfortable; yearly immunization for prevention	
common cold	caused by: rhinovirus, adenovirus spread by: droplet spread	discomfort but not as much as the flu, can cause other diseases	no cure	
chickenpox	caused by: varicella virus spread by: droplet spread	scratchy itchy spots on body, fever, headache	no cure, immunization for prevention; becoming less common because of recent immunization requirements	
smallpox	caused by: variola virus spread by: saliva droplet spread or direct contact with sores	small itchy spots, high fever, death	no cure, immunization for prevention; almost eradicated from the planet	
polio	caused by: poliomyelitis virus spread by: feces, hand to mouth	minor to severe muscle weakness and paralysis	no cure, immunization for prevention, almost eradicated from the planet	

Are Viruses Alive?

What makes something alive? When you were reading about bacteria, you read that bacteria reproduce, use energy, and grow. Viruses can also reproduce, use energy, and grow. However, unlike bacteria, viruses can only do this inside another cell.

To reproduce, a virus attaches itself to the cell membrane of a host cell. It then either injects (inserts) its genetic material into the host cell or enters the host cell. The genetic material contains all the information required to make new viruses. Inside the cell, the genetic material causes the host cell to produce the necessary components to make new viruses. The components made by the host cell then assemble to make new viruses. The new viruses are released from the host cell. The exact way that the new viruses are made and released varies from virus to virus. However, they all follow the sequence described. A virus must use the structures and metabolism of a host cell to reproduce.

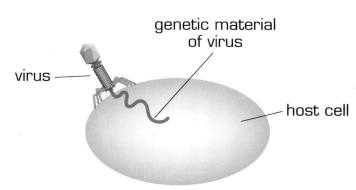

Scientists have learned a lot about viruses in the last 50 years. This has happened because new tools have made it possible to see the viruses. Very powerful microscopes, called electron microscopes, have made it possible for scientists to see viruses and identify how they work. Microscopes are so powerful now that scientists can see viruses as they move. This has made it possible to develop new ways to prevent viral illnesses.

Stop and Think

1. Explain one way that viruses can spread from person to person.

2. In what ways are viruses different from bacteria?

3. Describe the sequence of events that take place when a virus reproduces.

4. In your opinion, are viruses alive? Give reasons for your opinion.

What's the Point?

Viruses are one type of microbe that cause illness. There are many different types of viruses. Some viruses infect certain animals and others may infect certain plants. There is no cure for many viral illnesses. However, by learning about how viruses spread, you can prevent other people from getting sick.

2.9 Case Study

Investigating Smallpox

Bacteria and viruses can make you sick. The illnesses they cause can be very serious. Over the last 200 years, the number of bacterial and viral illnesses that can change your life has dropped dramatically. For example, 200 years ago most people got smallpox, almost as many as now get the flu. Unfortunately, many of those who were infected died of smallpox. Smallpox has been mostly **eradicated,** and you do not hear about people getting it anymore. Read about smallpox below. Use what you learn to fill in the *Communicable-Disease Information Table* with information about smallpox.

eradicated: wiped out.

Smallpox

Symptoms and Spread

Smallpox is a serious, contagious, and often fatal infectious disease. Smallpox is caused by the variola (vair-ee-OH-luh) virus. Humans are the only natural hosts of the variola virus. The symptoms include a rash and high fever. Smallpox is often recognized by the type of rash it causes.

The rash is in the form of small blisters. The blisters are filled with fluid and crusted over. This sounds like chickenpox, but the blisters look and feel different. The smallpox rash usually appears on exposed parts of the body: the face, arms, palms, lower legs, and soles of the feet. Other symptoms of smallpox include fever, headache, backache, and fatigue (tiredness).

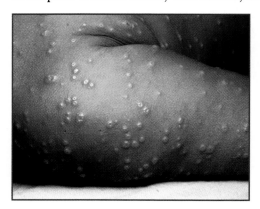

The "pox" means "spotted." It refers to bumps that appear on an infected person, as shown in the picture.

Generally, face-to-face contact is required to spread smallpox from one person to another. Smallpox also can be spread through direct contact with infected bodily fluids. If someone touches an object such as bedding or clothing that has smallpox on it, he or she may get infected. Sometimes, smallpox has been spread by viruses carried in the air. Smallpox is not known to be transmitted by insects or animals.

A person with smallpox becomes contagious when they get a fever. A person becomes more contagious when they get the rash. At this stage, the infected person is usually very sick and not able to move around in the community. The infected person is contagious until the last smallpox scab falls off.

Immunization

Have you ever heard of smallpox? Has anyone in your family had smallpox? You probably answered that no one in your family has had smallpox. This is because scientists have discovered how to prevent smallpox through **vaccination**. Edward Jenner was one of the first scientists to discover how to protect people from diseases through vaccination. He listened to stories about who got sick and who did not get sick and made sense of the reports he heard. He figured out how **immunization** might work based on these reports. These reports are called **case studies**.

The year was 1796. There were no cars, no telephones, and no electricity in houses. Edward Jenner heard that milkmaids, girls who milked cows, did not get smallpox. However, milkmaids did get a disease similar to smallpox from milking the cows. It was called cowpox. Unlike smallpox that killed many victims, the milkmaids only had a few blisters, felt a little tired, and had some aches.

vaccination: the process by which a substance that protects a person from a disease is given.

immunization: a medical treatment that helps protect you from disease.

case study: an observation of a person or group to use as a model.

This picture shows Jenner transferring fluid from a cowpox blister into a cut on a boy's arm.

He thought that maybe cowpox protected them. He decided to try an experiment. With the permission of the boy's father, Jenner gave an eight-year-old boy the cowpox virus that the milkmaids got. The boy had a few aches but did not get very sick. Later, Jenner exposed the boy to smallpox. The boy did not get sick. It seemed that the cowpox virus could protect the boy from smallpox. Several months later, Jenner again exposed the boy to smallpox disease. Again the boy stayed healthy. Jenner had found a way to prevent smallpox. The cowpox virus was acting as a **vaccine**.

vaccine: a substance that protects a person from a disease; from the Latin *vacca* for cow.

Trust in vaccinations came very slowly. But, little by little, people came to believe that the cowpox vaccine could prevent smallpox. Up until the 1970s, everyone was vaccinated against smallpox. This vaccination practice stopped because smallpox has been eradicated from the general population.

Before vaccines, parents in the United States could expect a whole variety of diseases to affect their children:

- *Polio*, a virus, would paralyze 10,000 children.

- *Rubella (German measles)*, a virus, would cause birth defects and mental retardation in as many as 20,000 newborns.

- *Measles*, a virus, would infect about 4 million children, killing 3,000.

- *Diphtheria*, a virus, would be one of the most common causes of death in school-aged children.

- *Haemophilus influenzae type b (Hib)*, a bacterial infection, would cause meningitis in 15,000 children, leaving many with permanent brain damage.

- *Pertussis (whooping cough)*, caused by a virus, would kill thousands of infants.

Just two generations ago, when your grandparents were children, they feared polio, rubella, and measles. Your parents also worried about measles, mumps, and rubella. Since Edward Jenner, vaccinations have been made for many different diseases. Vaccines have become commonplace in the last hundred years. In many cases, vaccines have eliminated diseases that killed or severely disabled people. Most children are given vaccinations as part of their health care in the first two years of their lives.

How Important Are Vaccinations?

Immunization has saved millions of lives in the last 200 years. In 1980, smallpox was declared eradicated. Polio has not been a problem for almost 50 years. These diseases can almost not be found. This means that people live longer, healthier, and better lives. Today, scientists continue to invent vaccinations for deadly illnesses. Some of those illnesses are bacterial (caused by bacteria) and some are viral (caused by viruses). Recently, scientists have developed a vaccine against a virus that can cause cancer in women.

What's the Point?

Scientists have developed many ways to combat some viral and bacterial diseases. Using vaccinations, they have eradicated some disabling and deadly diseases. Smallpox is one of those. Polio, chicken pox, measles, mumps, and rubella have also almost been eradicated. Vaccinations play an important role in maintaining people's health. Vaccinations are usually given to children early in their lives.

Scientists use new technology to see viruses and bacteria and to figure out how those microbes work. Then they are able to invent new ways of stopping them from working. These new scientists "stand on the shoulders of giants" like Edward Jenner. These scientists build on the work of Jenner, who used observations and case studies to figure out that immunization was possible.

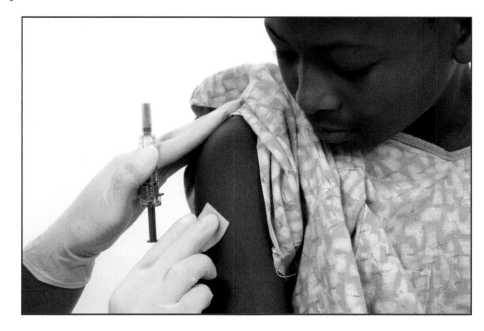

GOOD FRIENDS and GERMS

PBIS

2.10 Case Studies

Disease Overviews

Edward Jenner used case studies to learn more about the viral illness smallpox. On the following pages, you will read case studies about five people with different diseases. You will learn a lot about the diseases and the people who have them.

Reading some of the case studies might be difficult because they contain a lot of medical terms. For example, in the first reading, about chickenpox, the word varicella-zoster is used. This is the medical term for chickenpox. One way to read this type of word is to pronounce it the best you can. Get the main idea of the word, and then keep reading. As long as you get the main ideas of the case study, you do not need to pronounce or remember every word exactly.

Procedure

The case studies that follow give you information about five different diseases.

1. Read the case study (or case studies) your teacher assigns you.

2. As you are reading the case study, look for the information that helps you answer the questions in the Communicable-*Disease Information Table*. You will need to know
    • the name of the disease,

    • whether it is caused by a bacteria or a virus,

    • how it spreads,

    • how it affects your body systems,

    • the symptoms, and

    • how it is treated.

In the column about how it is treated, tell whether the disease has a vaccine. A sample entry into the table is shown. Notice the questions you should try to answer in each column.

Project-Based Inquiry Science

Be a Scientist

Case Study: Chickenpox

India is two years old. She woke up this January morning with a low fever, and she feels terrible. She also has spots all over her. The spots are red and quite large and are mostly on her belly, face, chest, and back. They itch a lot. India is finding it hard not to scratch the spots. The last time her mother looked at India's belly and chest, there were more.

India likes to play with her three-year-old cousin. Last week her cousin could not play with her because he had the chickenpox. Now India's mom is sure she also has chickenpox.

India has two older brothers. India's mother is worried about them now.

Communicable Disease Information Table						2.10.1

Name: _____ Date: _____

Name of disease	Cause (bacterial or viral)	How does it spread?	Which body systems does it affect?	What are the symptoms?	How is it treated? Is there a vaccine?	Other notes
Write the scientific name and the common name. People usually say that they have "the flu," which is a shorter way to say that they have "influenza."	Many diseases are caused by either a virus or a bacterium.	Some diseases are spread through blood, body fluids (like saliva), air, contaminated food, and animal bites.	Some diseases affect several systems. List the main one(s).	The symptoms are what help you to determine that you have a particular disease.	"Treat the symptoms" means doing something so the symptoms will not be as bad. Sometimes, you can treat a disease, sometimes you can only treat the symptoms.	Use this column to write anything you think is interesting or important to remember about a particular disease.
common cold	viral	Body fluids in the air or on surfaces—blowing your nose, touching your hands to other people or objects, sneezing or coughing.	respiratory	cough, body aches, congestion	It can't be cured. You can only relieve the symptoms to feel better, using medicine to reduce the fever and to stop the runny nose, and cough.	

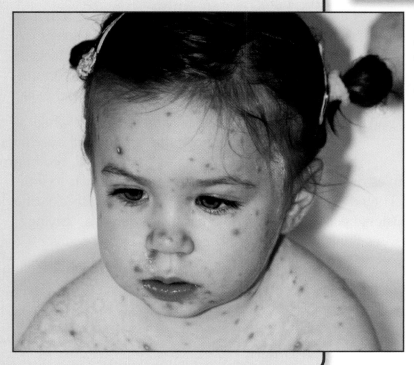

Chickenpox was once a common childhood disease. The number of cases of chickenpox in the United States has been reduced since a vaccine was introduced in 1995.

Learn about Chickenpox

Chickenpox is a common disease caused by a virus called the varicella-zoster virus (VZV). Varicella-zoster virus spreads in the air through coughs or sneezes or through contact with fluid from inside the chickenpox blisters.

Chickenpox, which occurs most often in late winter and early spring, is a communicable disease. It can be spread very easily. Chickenpox is very contagious. If your brother gets chickenpox and you have not had it yet, there is an 80%–90% chance that you will get it too.

Although it is more common in persons under the age of 15, anyone can get chickenpox. A person has only one episode of chickenpox in his or her lifetime.

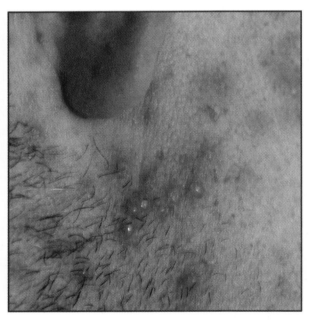

One of the symptoms of chickenpox is an itchy rash.

If you have a younger brother or sister, they may never get chickenpox. For the last few years, the chickenpox vaccine has been required of all children before they go to school. The vaccine gives you some of the chickenpox virus, but not enough to get you sick. Your body thinks you have had chickenpox and builds up defenses. You cannot get chickenpox again.

One of the symptoms of chickenpox is a red, itchy rash on the skin. The chickenpox rash usually appears first on the abdomen or back and face. Then it spreads to almost everywhere else on the body, including the scalp, mouth, nose, and ears.

Some people can have only a small rash, while others have a serious rash that spreads all over their body. The rash may begin as many small, red bumps that look like pimples or insect bites. These bumps are about two to four millimeters wide. They develop into thin-walled blisters filled with clear fluid. The fluid in the blisters becomes cloudy. The blister wall breaks, leaving open sores. The sores finally crust over to become dry, brown scabs.

Some infected people have a fever, abdominal pain, or a vague sick feeling a day or two before the rash appears. These symptoms may last for a few days. The fever stays in the range of 37.7°C to 38.8°C (100°F to 102°F). Younger children often have milder symptoms and fewer blisters than older children or adults.

Be a Scientist

Case Study: Lyme Disease

Last week, Shane took a walk in the woods with his dog. Shane lives in Maine where there are lots of fields and forests. Shane loves to take frequent walks to look at the wildlife and get some fresh air. This morning Shane woke up with a sore spot on his leg. It looks like a **bull's-eye**. He also has a fever and headache and his muscles hurt. Shane is going to stay home from school today.

Shane's dad called the doctor, who said to look very carefully at the center of the bull's-eye. Shane's dad used a magnifying glass to look at his leg. In the center of the bull's-eye, Shane's dad found a small insect. He carefully pulled it out, making sure that he got the whole insect.

The doctor prescribed some antibiotics for Shane, because Shane's illness is caused by bacteria. The medicine will help kill the bacteria that caused Shane's disease. It may take a long time for Shane to feel better.

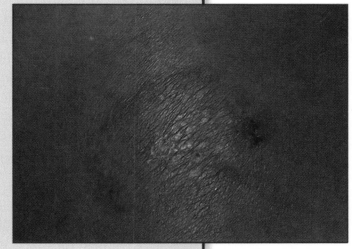

The first symptom of Lyme disease is a bull's-eye rash.

bull's-eye: a circular spot, usually black, at the center of a target. The bull's-eye is surrounded by concentric circles.

Learn about Lyme Disease

Lyme disease is caused by bacteria that are found in some ticks. The ticks live in wooded areas and fields. This is one reason why it is so important to check yourself for ticks after being outside, even in your backyard or neighborhood park.

Within one to two weeks of being bitten by the tick, infected people may have a "bull's-eye" rash. They may also have a fever, a headache, and muscle or joint pain. Some people have Lyme disease and do not have any early symptoms. Other people have a fever and other "flu-like" symptoms without a rash.

Not all ticks carry the bacteria that cause Lyme disease. A tick can become infected with the bacteria after it bites and feeds on an infected animal, often a deer. The bacteria live inside the tick and are passed into the animal that is bitten. Sometimes that bitten animal is a person.

If Lyme disease is left untreated, after several days or weeks, bacteria may spread throughout the body of an infected person. Then there may be many more symptoms. An infected person may get rashes on other parts of the body. Symptoms also may include pain that seems to move from joint to joint. There may be signs of inflammation of the heart or nerves.

If the disease is still not treated, infected people can get additional symptoms. These can include swelling and pain in major joints, or mental changes. Some symptoms might not show up until months after getting infected. The treatment usually involves using different types of antibiotics.

While most of the cases of Lyme disease in the United States are in the Northeastern part of the United States, this disease is found all over the world and is a concern across the United States. Montana is the only state that has never reported a case of Lyme disease. During spring and summer, you will often hear news stories about Lyme disease. It is important to be careful if you live in an area that has deer ticks, because they may carry the bacteria that cause Lyme disease.

Adult female deer ticks like the one in this picture may carry the bacteria for Lyme disease.

Be a Scientist

Case Study: *E. coli* Infections

It was August and September, 2006, and calls began to come into doctors' offices and hospitals around the country. People were complaining of diarrhea, fever, cramps, and vomiting. Most people felt very sick for several hours and missed school or work. Many people were hospitalized. Over the next few days, there were more calls from people all over the country. Officials counted calls from 25 states. Eventually, by the first of October, 199 people had gotten sick. Three people had died.

The calls from sick people began arriving about August 19. Investigators from the Food and Drug Administration (FDA) spoke with many of the sick people. Thinking that the illnesses were being caused by something in food, the FDA used a survey to identify what foods the people had eaten. The survey had hundreds of foods on it. People told the FDA what they had eaten in the last several days. When all the data were analyzed, spinach came out as a food most people had eaten three to four days before they got sick. The FDA began to look for contaminated spinach. They sent out the press release shown on the next page.

FDA officials also warned that *E. coli (Escherichia coli)* could be spread among people who did not maintain good hygiene. *E. coli* is found in the stools of people who are infected. This is especially a problem for people who take care of children who are not toilet trained. The FDA reminded people to wash their hands well after using the bathroom and after taking care of small children.

FDA Warning on Serious Foodborne E.coli O157:H7 Outbreak

One Death and Multiple Hospitalizations in Several States

The U.S. Food and Drug Administration (FDA) is issuing an alert to consumers about an outbreak of E. coli O157:H7 in multiple states that may be associated with the consumption of produce. To date, preliminary epidemiological evidence suggests that bagged fresh spinach may be a possible cause of this outbreak.

Based on the current information, FDA advises that consumers not eat bagged fresh spinach at this time. Individuals who believe they may have experienced symptoms of illness after consuming bagged spinach are urged to contact their healthcare provider.

"Given the severity of this illness and the seriousness of the outbreak, FDA believes that a warning to consumers is needed. We are working closely with the U.S. Centers for Disease Control and Prevention (CDC) and state and local agencies to determine the cause and scope of the problem," said Dr. Robert Brackett, Director of FDA's Center for Food Safety and Applied Nutrition (CFSAN).

E. coli O157:H7 causes diarrhea, often with bloody stools. Although most healthy adults can recover completely within a week, some people can develop a form of kidney failure called Hemolytic Uremic Syndrome (HUS). HUS is most likely to occur in young children and the elderly. The condition can lead to serious kidney damage and even death. To date, 50 cases of illness have been reported to the Centers for Disease Control and Prevention, including eight cases of HUS and one death.

At this time, the investigation is ongoing and states that have reported illnesses to date include: Connecticut, Idaho, Indiana, Michigan, New Mexico, Oregon, Utah, and Wisconsin.

FDA will keep consumers informed of the investigation as more information becomes available.

	Number of persons with *E.coli* infection
Arizona (AZ)	8
California (CA)	2
Colorado (CO)	1
Connecticut (CT)	3
Idaho (ID)	7
Illinois (IL)	2
Indiana (IN)	10
Kentucky (KY)	8
Maryland (MD)	3
Maine (ME)	3
Michigan (MI)	4
Minnesota (MN)	2
Nebraska (NE)	11
New Mexico (NM)	5
Nevada (NV)	2
New York (NY)	11
Ohio (OH)	25
Pennsylvania (PA)	10
Tennessee (TN)	1
Utah (UT)	19
Virginia (VA)	2
Washington (WA)	3
West Virginia (WV)	1
Wisconsin (WI)	45
Wyoming (WY)	1

Learn about *E. coli*

Escherichia coli is the formal scientific name for *E. coli*. It is pronounced as esh-er-ick-ee-eh cole-eye, but it is usually referred to as *E. coli*. Symptoms of *E. coli* include watery or bloody diarrhea, fever, abdominal cramps, nausea, and vomiting. Illness may be mild or severe. Very young children and very old or sick people are more likely to have severe symptoms, sometimes including kidney failure and death.

E. coli is a common cause of foodborne illness. An estimated 73,000 cases of *E. coli* infection and approximately 60 deaths occur in the United States each year. Most illness has been associated with eating undercooked, contaminated ground beef. Person-to-person contact in families and childcare centers is also an important mode of transmission. Infection can also occur after drinking milk that has not been **pasteurized**. A person can also be infected after swimming in or drinking sewage-contaminated water.

There are several things around your house you can do to prevent *E. coli* infection. For example, it is important not to use the same cutting board for raw meat and other foods. If you do, you can transfer bacteria that were on your meat to your vegetables. You might not get sick from the meat, because you cook it, but you can get sick from the vegetables, because you may not cook them. You also should use a different plate to carry uncooked meat out to the barbeque grill and to carry the cooked meat back inside.

Uncooked spinach can sometimes harbor E. coli.

Be a Scientist

Case Study: Mononucleosis

A fifteen-year-old high-school student named Ally has been complaining about feeling ill for about three days. Ally's mom has decided it is time to take Ally to the doctor. The doctor examines Ally and asks her to describe her symptoms. Ally says she feels very tired and her neck is sore and stiff. The doctor feels her neck in the back and on the sides. She asks Ally to bend her neck forward. Then the doctor asks Ally to lie down, and she presses on her belly. She notices that Ally cannot bend her neck as far as she is supposed to and that the glands in the sides of her neck are lumpy and swollen. When the doctor looks at Ally's throat, she notices that it is very swollen and red. There is no drainage from her nose (the drainage might indicate a cold). When the doctor pressed Ally's belly, she could feel that Ally's spleen was slightly swollen. Ally did not have any skin rash and did not have a fever. No one Ally knows has been sick lately.

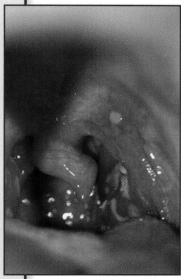

One symptom of mono is white patches on the back of the throat.

Learn about Mononucleosis

Infectious mononucleosis (mah-noh-new-klee-oh-sis) is often called mono. It is sometimes known as the kissing disease. It can spread through kissing, but it can also spread through coughing, sneezing, or sharing a glass or cup. The Epstein-Barr virus causes mononucleosis. However, similar signs and symptoms are sometimes caused by cytomegalovirus (CTV). Full-blown mono is most common in adolescents and young adults. Signs of mono can include fever, sore throat, headaches, white patches on the back of your throat, swollen glands in your neck, feeling tired, headaches, sore muscles, and not feeling hungry. A sick person does not need to have all of these symptoms to have mono.

The virus that causes mononucleosis is transmitted from one person to another through the saliva of an infected person. The virus can continue to be excreted in saliva and **sputum** (matter that you cough up) for as long as six months after the infected person has gotten better. This is one reason why it is important not to share things, like straws or lip gloss, at school with your friends. A person does not need to feel sick to make other people sick with this disease.

Mono makes you feel tired and achy, but it usually is not very serious. The virus remains in your body for life. Most people have been exposed to the Epstein-Barr virus by the time they are 35 years old. Because they have had the virus, they are immune and cannot get mononucleosis again.

pasteurized: to heat food to a temperature that is high enough to kill most harmful bacteria.

sputum: matter that is coughed up and mixed with saliva.

Be a Scientist

Case Study: Pneumonia

An elderly man reports to his nurse that he has had a bad cold for about ten days. He has been coughing and has had a high fever with shaking and chills. He tells the nurse that he does not understand why the cold will not go away. He has been coughing a lot, and the coughing produces sputum that is rust-colored. This morning when he woke up, he had a terrible shooting pain in the right side of his chest. He is also having difficulty breathing.

The nurse collects some of the sputum, gives the man oxygen through a tube in his nose, and admits him to the hospital. The man begins a treatment of antibiotics. The nurse sends the sputum to the laboratory to have it tested for bacteria. When all the data comes back from the lab, the doctor tells the nurse that the man has bacterial pneumonia. He probably caught it by breathing in the bacteria coughed into the air by someone else.

Learn about Pneumonia

pus: a white, or slightly yellow or green substance that your body develops in response to an infection; it is made up of dead skin, white blood cells (that fight infection), and some bacteria.

Pneumonia is a serious infection or inflammation of the lungs. Air sacs in the lungs fill with **pus** and other liquid so that oxygen has trouble getting into the blood. If there is too little oxygen in the blood, a person's body cells cannot work properly. Because of this, and because infection can spread throughout the body, pneumonia can cause death.

Until 1936, pneumonia was the number one cause of death in the United States. Since then, antibiotics are used so that fewer people die from pneumonia. There are 30 different causes of pneumonia. The most common causes are bacteria and viruses. Some of the similarities and differences are described in the following sections.

Bacterial Pneumonia

Pneumonia bacteria are present in some healthy throats. When body defenses are weakened in some way, the bacteria can multiply and cause serious damage. Body defenses are weakened when a person is very young or old, when the person is not in good health, when the person is in a weakened condition like just after having surgery, or when the person has impaired immunity. When a person's defenses are weakened (sometimes people say their resistance is lowered), bacteria work their way into the lungs and inflame the air sacs.

Sometimes the symptoms of bacterial pneumonia happen gradually. Sometimes they happen suddenly. A person may experience shaking chills, chattering teeth, severe chest pain, and a cough that produces rust-colored or greenish sputum. Body temperature may rise as high as 40.5°C (105°F). The person sweats profusely, and breathing and pulse rate increase rapidly. Lips and nail beds may have a bluish color due to lack of oxygen in the blood. In the most serious cases, the person may be confused or delirious. Streptococcus pneumoniae is the most common cause of bacterial pneumonia. It is one form of pneumonia for which a vaccine is available.

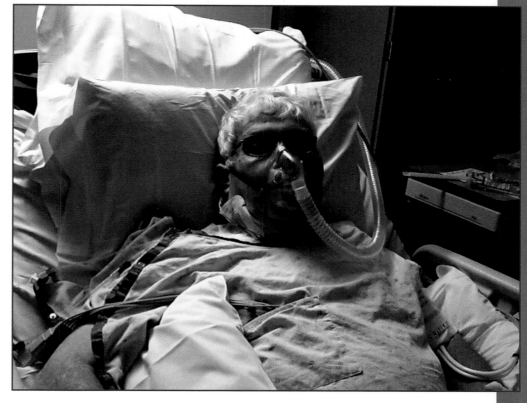

The symptoms of pneumonia can range from mild to serious, depending on the age of the person and the type of pneumonia.

Viral Pneumonia

Half of all pneumonias are believed to be caused by viruses. More and more viruses are being identified as the cause of respiratory infection. Some produce pneumonia, especially in children. Most of these pneumonias are not serious and last a short time. The virus invades the lungs and multiplies. There are almost no physical signs of lung tissue becoming filled with fluid. Many people who get viral pneumonia are those who have breathing conditions or who are pregnant.

The initial symptoms of viral pneumonia are fever, a dry cough, headache, muscle pain, and weakness. Within 12 to 36 hours, there is increasing breathlessness. The cough becomes worse and produces a small amount of mucus. There is a high fever, and a person's lips may look bluish. In extreme cases, the patient has a desperate need for air and extreme breathlessness. Sometimes viral pneumonia is complicated by an invasion of bacteria, and the patient has two types of pneumonia.

Communicate

Share Your Case *Study*

Each of the case studies in this section presented a different example of a communicable disease. Since each group read only one case study, it is important that the information is shared with the class. Use the information you recorded on your *Communicable-Disease Information Table* to make presentations to your class about what you learned from the case study you read. Your classmates will be relying on you to learn about different examples of communicable diseases. It is important for you to include the following information:

- the name of the disease
- whether it is caused by a bacteria or a virus
- how it spreads
- how it affects your body systems
- the symptoms
- how it is treated

As a presenter, it is important that your presentation include all of the above points. As you listen to your classmates' presentations, it is just as important that you hear and understand this information. If you do not understand something, or if you think presenters left out something important, ask questions. Be careful to ask your questions respectfully, and do not interrupt your classmates' presentations.

What's the Point?

Learning about different diseases helps you understand how you can get sick. It is interesting to see how easily some diseases are spread from one person to another.

The chickenpox virus can infect others when an infected person coughs or sneezes. The bacterium that causes Lyme disease can spread to others by the bite of a deer tick. *E. coli* is a bacterium usually found in contaminated food and is responsible for causing gastrointestinal infections. Mononucleosis, a viral disease, can be spread through the saliva of an infected person for up to six months after the person is feeling better. Bacterial pneumonia, which can be treated with antibiotics, usually occurs in people who are already sick with something else. Viral pneumonia is usually less severe than bacterial pneumonia.

Learning Set 2

Back to the Big Question

How can you prevent your good friends from getting sick?

The *Big Question* for this Unit is *How can you prevent your good friends from getting sick?* To help you answer this question, you looked at cells. Your body is made up of cells. To understand what happens to your body when you get sick, you need to know about what makes up a cell and how a cell functions.

You know that bacteria and viruses can make you sick and cause disease in your body. You also know that some bacteria are needed by your body to keep you healthy. Bacteria and viruses are both microbes (microorganisms). You cannot see them without a microscope. Bacteria are single-celled organisms. You need to know how a cell functions to understand how bacteria function.

Viruses are even smaller than bacteria. You need to use a special type of microscope to see them. There is no cure for diseases caused by viruses. Preventing the viral disease from spreading is the best way to stay healthy.

Explain

You know so much more now about how you can get sick from communicable diseases. It is time to revise your explanations of how you can get sick from communicable diseases. Begin by revisiting the claims you made earlier. Work on making them more specific. You might want to add specifics to your claim about how you can get sick from a communicable disease by touching everyday objects or by eating everyday foods. What you've read about specific communicable diseases can help you make your claims more specific.

Create Your Explanation

Name:_____ Date:_____

Use this page to explain the lesson of your recent investigations.

Write a brief summary of the results from your investigation. You will use this summary to help you write your Explanation.

Claim – a statement of what you understand or a conclusion that you have reached from an investigation or a set of investigations.

Evidence – data collected during investigations and trends in that data.

Science knowledge – knowledge about how things work. You may have learned this through reading, talking to an expert, discussion, or other experiences.

Write your Explanation using the *Claim*, *Evidence* and *Science knowledge*.

Record your new more specific claim on a new *Create Your Explanation* page. Then record evidence and science knowledge that support your claim, and write a statement that connects your claim and your evidence and science knowledge. Make sure your statement discusses the role bacteria and viruses play in getting you sick and how they are passed between people. Remember that a good explanation can convince someone else of your claim. If your statement doesn't seem convincing, revise your claim so your evidence and the science knowledge you know support it. Be prepared to share your claims and explanations with the class.

Communicate

Share Your Explanation

Share your explanation with the class. Make sure to be clear about your claim, the evidence and science knowledge that support it, and your explanation statement. As you listen to the explanations of others, make sure you understand them completely and agree with them. If you do not understand something, or if someone has not been clear, ask them to add more detail to what they are saying. If you think someone is wrong, respectfully disagree, being sure to give evidence to support your disagreement.

Revise Your Explanation

After everyone has presented, as a class, develop an explanation about how you get sick from communicable diseases that everybody agrees with.

Recommend

Using your explanations as a guide, revisit your recommendations, and develop more specific recommendations about staying well. Your recommendations should answer the same two questions you answered in Section 1.3.

- How can I keep from getting sick with a communicable disease?

- How can I prevent my friends from getting sick with a communicable disease?

Remember that you now know much more about which everyday objects have a lot of germs on them and the locations where you are especially likely to come into contact with germs. For each recommendation you made previously, decide if you still think it is a good recommendation. If you think it is but you can state it better, restate your recommendation. Then add any new evidence or science knowledge that supports it. Revise your explanation statement to make it more complete.

You might also want to make new recommendations. Use a new *Create your Explanation* page for each. State your claim as a recommendation and add evidence and science knowledge that support it. Then develop a statement that connects your recommendation to your evidence and science knowledge. Be prepared to share your recommendations with the class and to support them with evidence and science knowledge.

Communicate

Share Your Recommendation

Share your recommendations with the class. When you present, be clear about the evidence, science knowledge, and explanation that support each of your recommendations. As you listen, be sure that you understand each recommendation your classmates make and that you are convinced the recommendation is good. If the evidence or science knowledge does not support it well, respectfully disagree with the presenter and report the evidence or science knowledge that tells you the recommendation is not convincing.

GOOD FRIENDS and GERMS

Conference

With your group, identify three things you have read about in this *Learning Set* that you think should be added to the *Project Board* and the evidence that supports each.

Update the *Project Board*

The *What are we learning?* column on the *Project Board* helps you pull together what you have learned. Share the things you think are important to add to the *Project Board* and your evidence for each. As a class, decide which of these things belong in the *What are we learning?* column. Remember to include evidence in the *What is our evidence?* column. You must fit the pieces together to help you address the challenge. Your *Big Question* was *How can you prevent your good friends from getting sick?* The last column, *What does it mean for the challenge or question?*, is the place to record how learning about bacteria and viruses can help you answer the *Big Question*. Remember to keep a record of what was added to the class *Project Board* on your own *Project Board* page.

How can you prevent good friends from getting sick?				
What do we think we know?	What do we need to investigate?	What are we learning?	What is our evidence?	What does it mean for the challenge or question?

Project-Based Inquiry Science

Learning Set 3

What Happens to You When You Get Sick?

When you are sick, your body does not feel normal. Your throat might hurt, your nose might be stuffy, and it might be hard to breathe. Different illnesses make you feel bad in different ways. Recall that these indications that you are sick are called *symptoms*. Doctors and nurses use symptoms as clues to determine the disease you might have. Symptoms are connected to body systems.

You first modeled how a disease could spread. Then you learned that bacteria and viruses are microbes that can make you sick and cause disease. In this *Learning Set*, you will investigate some of the body systems that are affected by communicable diseases. You will also look at the immune system, which helps you stay healthy. You will see how microbes can move within your body and how the symptoms you feel are caused. This will help you answer the question, *What happens to you when you get sick?*

The amazing thing about your body is how all the parts work together. You can walk and talk and move because of the ways the parts of your body work. Air moves into your lungs. The oxygen in the air then moves into your blood. The blood moves through your body. It brings the oxygen to each cell where it is needed. The way your body works also allows you to spread germs and to transmit diseases to others.

In this *Learning Set*, you will be learning more about each of your body systems. As you build models of the body systems, you will think about how the parts of the systems work together to keep you healthy. You will also learn about the role each system plays in spreading disease. Learning how your body works will help you stay healthier and help you avoid making others sick.

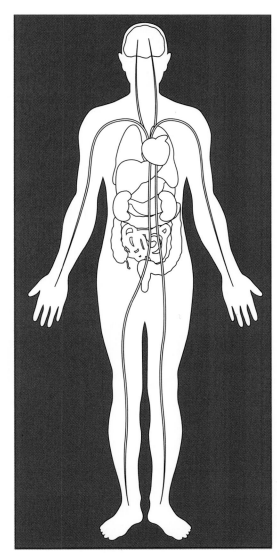

The human body is made up of many systems, including the respiratory, circulatory, and digestive systems.

3.1 Understand the Question

Thinking about What Happens to You When You Get Sick

The question for this *Learning Set* is *What happens to you when you get sick?* To help you answer this question, you will be learning about body systems. It is a good idea to think about what you know about systems, and what you know about the systems in your body. It is also important to think about what you are unsure about and what you would like to investigate.

Get Started

A system is like a team of players. Each player on the team has a job to do. One player on the team may be very important and talented. However, without the other players, he or she cannot get the job done.

Think about some systems that you already know. For example, you may know about the transit system in your community (buses, subway, trains). You may know about sound systems, emission systems on cars, alarm systems, or irrigation systems for a farm. Each system has parts. Each of the parts has a job, or function. In a system, the parts work together to do a job. If you think of a bicycle as a system, the pedals, chain, seat, handlebars, and you work together to make a bicycle move.

What do you already know about your body systems? You probably know the names of some organs and some body systems. Record what you know about the names of organs in each of these body systems:

- the respiratory system,
- the circulatory system, and
- the digestive system.

Project-Based Inquiry Science

(Here are some examples of organs to get you started: heart, intestines, skin, brain, lungs, stomach, veins, esophagus, and trachea. Don't worry if you don't know all of these. During the next few days, you will be learning more about each of these systems and organs.)

On your own, think of several questions that might help you answer the question for this *Learning Set*. Develop two questions that might help you understand body systems better.

In your group, share your ideas about what you think you know about body systems. Listen carefully to all the ideas presented. Then share your two questions with your small group. Carefully consider each question and decide if it meets the criteria for a good question. With your group, refine the questions that do not meet the criteria. Choose the two most interesting questions to share now with the class. Give your teacher the rest of the questions so they might be used later.

Update the *Project Board*

Share your group's ideas about body systems with the class. You can record these ideas in the *What do we think we know?* column. Next, share your group's two questions with your class. Your teacher will add your questions to the *Project Board*. In this *Learning Set*, you will work to answer some of these questions.

How can you prevent good friends from getting sick?				
What do we think we know?	What do we need to investigate?	What are we learning?	What is our evidence?	What does it mean for the challenge or question?

3.2 Read

From Cells to Systems

You learned that the building blocks of all living things are cells. Unlike the bacteria you read about that can be single-celled organisms, you are made up of trillions of cells. The types of cells in your body are varied in size, shape, and how they function. (Function means how they work or the job they do.) Each different type of cell is like a "specialist." Cells in your body are specialized to do different things. However, they must work together to keep your body functioning properly. In this section, you will read about how the cells in your body are organized.

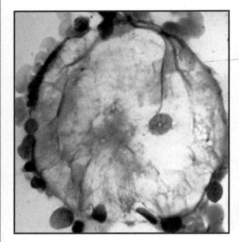
A marrow fat cell.

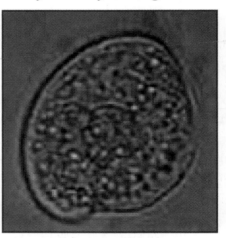
An epithelial cell.

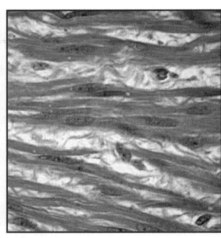
A type of muscle tissue made up of muscle cells.

tissues: groups of cells that are similar in structure and function.

contract: shorten.

organs: structures that have a specific function and are made up of different tissues.

organ systems: groups of organs that have related functions.

How Are Cells Organized?

The cells in your body have many shapes and functions. Cells that are similar in structure and function are called **tissues**. Muscle is an example of one type of tissue. It is made of muscle cells. Muscle tissue is able to **contract**. That is what makes parts of your body able to move.

Tissues are organized into larger structures called **organs**. Organs are groups of different tissues that have a particular job. For example, your heart is an organ made up of muscle, nerve, fat, and other tissues. The nerve tissue gets the muscle tissue in the heart to contract. That makes blood move through the body.

Organs are grouped together in **organ systems**. The organs in a system have related functions. The circulatory system includes the heart, veins, and arteries. They work together to get the job done.

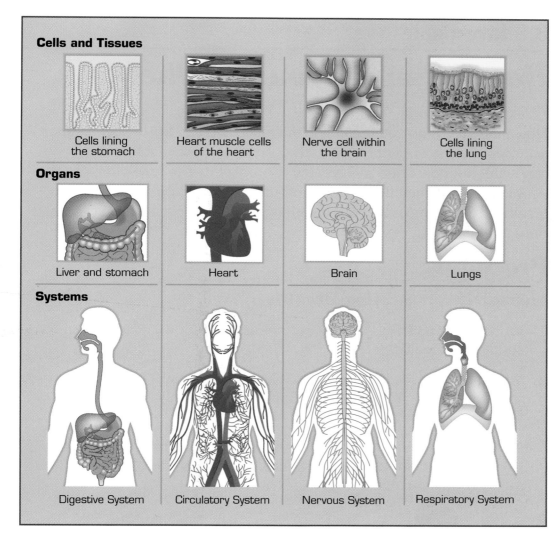

Cells and Tissues

Cells lining the stomach | Heart muscle cells of the heart | Nerve cell within the brain | Cells lining the lung

Organs

Liver and stomach | Heart | Brain | Lungs

Systems

Digestive System | Circulatory System | Nervous System | Respiratory System

Stop and Think

The levels of organization in a multicellular organism are cells, tissues, organs, and organ systems. Describe each of these levels of organization.

What's the Point?

Your body is made of organ systems. Every system has a different job to do. Systems are made of parts. The parts are organs, tissues, and cells. Every part in a system does a job to make the whole system work. In your body, each system does a different job to keep you healthy. When you know more about how your body works, you can stay healthier. Learning about your body systems will make it easier to understand your body and how to keep it healthy.

3.3 Investigate

How Often Do You Breathe?

How many times do you breathe in a minute? Do you know how different activities affect how often you breathe? Try this investigation to see if you can figure out how often you breathe. Compare that to how often other people breathe. Then think about how different activities affect how you breathe. You will measure your breathing two times, once while at rest and once after exercise.

Procedure

Measure Your Resting Breaths

Materials
- **stopwatch**
- **Breathing Rate data pages**

1. Work with a partner. Let your partner use the stopwatch while you count your breaths.

2. Your partner starts the stopwatch. Count your breaths until your partner tells you to stop. Your partner will tell you to stop after 6 seconds. Do this two more times.

3. Multiply your number of breaths by 10 to determine how many times you breathe in one minute (6 seconds times 10 equals 60 seconds or one minute).

4. Switch roles and repeat Steps 1 to 3.

Measure Your Resting Breaths after Exercise

You are going to repeat the steps you just did, but with exercise as part of the investigation. Rather than measuring your resting breaths, you are going to measure how many times per minute you breathe right after you do a little exercise.

5. Work with your partner again. Stand near your desk and make sure you are a safe distance from everyone else. Jog in place for 30 seconds.

6. Sit down and begin counting your breaths immediately. Stop when your partner tells you that 6 seconds have passed.

Record your data. Begin counting your second trial immediately after the first, and your third right after the second. Record all three trials.

7. Multiply your numbers by 10 to determine how many times you breathe in one minute.

8. Switch roles and repeat Steps 4 to 7.

Recording Your Data

Scientists always keep track of their data as they collect it. Use a table like the one below to track the data you collect. Use one table for yourself and one for your partner. Complete the tables so you have a record of your data.

	My Breathing Rate Data			3.3

Name: _____ Date: _____

Trials	My number of breaths in 6 s without exercise	My number of breaths in 1 min without exercise (breaths in 6 s x 10)	My number of breaths in 6 s with exercise	My number of breaths in 1 min with exercise (breaths in 6 s x 10)
1		x 10 =		x 10 =
2		x 10 =		x 10 =
3		x 10 =		x 10 =
Average number of breaths per minute for all trials	Add together the number of breaths per minute for each trial and divide by 3			

Stop and Think

1. An average person breathes 15 to 25 times per minute. This is called a range. Does your number of resting breaths fall within that range? If it does not, why do you think it is different? Does your partner's number of breaths fall in that range? If not, why do you think it didn't?

2. How did your results (yours and your partner's) compare for resting and exercising?

3. Were your resting or exercising results higher or lower?

4. Why do you think they were different?

Communicate Your Results

Share your pair's results with the class. Create a class table that includes the average breaths per minute for everyone. The table should look like the one shown below. You might use a similar table to organize all the data for yourself.

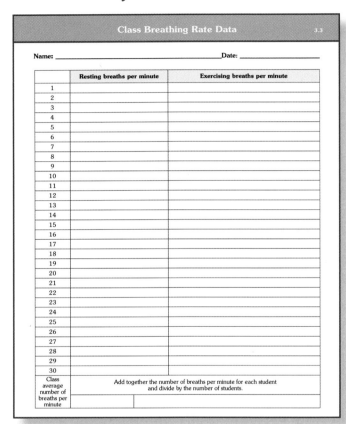

Analyze Your Data

Now that you see everyone's data, you can compare your own information to the class information. When scientists have collected a set of data like this, they analyze it to see what information is important to answering their question. Your question is, *How often do you breathe?* You answered that question for yourself and your partner, but a bigger set of data helps to answer the question in a larger group or population.

1. What was the range of average number of breaths per minute in your class at rest and after exercising?

2. What was the class average for the number of breaths per minute at rest and after exercising? Your teacher will help you use a calculator to figure that out.

3. Did the average for resting fall within the range of 15 to 25 breaths per minute? Was it at the high end, closer to 25 breaths, or at the low end, closer to 15 breaths?

4. How did your at-rest and after-exercising breaths per minute compare to other people's?

5. How reliable do you think the data are? Why?

Reflect

Answer the following questions. Be prepared to discuss your answers with the class.

1. The average resting rate was probably lower than the average exercising rate. Why do you think that is?

2. What was the high number for resting breathing rate? What was the low number? What was the class average number for these rates? Which is the best number to use when you want to know about the population, the high, the low, or the average number of breaths? Why did you choose that number?

3. Why do you think you would need to breathe more often when you are exercising than when you are resting?

What's the Point?

In this investigation, you measured how many times you breathe in one minute when you are sitting still. You compared your at-rest breathing rate to the class average. Then you compared your class average to the average for the general population. Next, you compared the at-rest breathing rate to the after-exercising breathing rate. Your exercising breathing rate was faster than your resting rate. You took more breaths per minute after you exercised than when you were at rest.

Breathing rate increases during exercise because the body needs more oxygen than when at rest.

GOOD FRIENDS and GERMS

3.4 Investigate

How Does the Respiratory System Work?

When you are sick, a medical person may ask you to describe how you feel. He or she is asking you to help identify the symptoms of your disease. If you have a cold, you may say that your nose is stuffy. You may also say that you are coughing. The medical person will think about which diseases affect your body that way. He or she will think about diseases of the **respiratory system**. That is because your nostrils, which are stuffed up, are part of that system. That person may listen to your lungs because they are also part of your respiratory system.

respiratory system: the organ system that delivers oxygen to the blood and removes carbon dioxide, a waste gas, from the blood. It includes the nostrils, trachea, and lungs.

trachea (windpipe): a tube that carries air to the lungs.

Like all other systems in the body, the respiratory system is made up of particular organs that have particular functions. Some of the organs of the respiratory system include the nose, the **trachea** or windpipe, and the lungs. The trachea is the tube that takes air from your nostrils to your lungs. The function (job) of the respiratory system is to deliver oxygen to the body and remove carbon dioxide from the body. You need oxygen to live. Carbon dioxide is a waste gas that comes from normal body processes.

By listening to the sounds your lungs make as you breathe, a medical person can assess the health of your respiratory system.

It is hard to imagine how your respiratory system works because you cannot see it. Because you cannot see it, building a model is a good idea. You can build a model of the respiratory system with very common materials. When you build it, you will be able to see how the organs of the respiratory system work together to help you breathe.

Build and Run Your Model

1. First, explore what happens when you breathe in and out. Place one hand on your ribs. Place your other hand just under your ribs. Take four or five deep breaths. Observe what happens under each of your hands as you breathe.

2. Answer the following questions:
   a) What do you feel under each of your hands?

   b) What do your lungs do under your hand when you breathe?

   c) Your other hand is just under your ribs. How does this part of you move when you breathe?

3. Using the materials, build your model to match the diagram. Attach a balloon to one end of the straw. Make sure that the fit is tight. You may wish to use a rubber band to make a tight fit.

4. Place the "lung and windpipe" (balloon and part of the straw) into the mouth of the bottle. Secure the straw with a lump of clay.

5. Stretch the plastic wrap tightly over the bottom of the bottle. Secure it with a rubber band.

6. After you have built your model, run your model and answer the following questions. Be prepared to share your answers with the class.
   a) Pull on the plastic wrap at the bottom of the model. What happens to the balloon inside the model? How is this like inhaling?

   b) Push the plastic wrap up at the bottom of the model. What happens to the balloon inside the model? How is this like exhaling?

Materials
- **straw**
- **balloon**
- **2 rubber bands**
- **2-L plastic bottle with bottom cut off**
- **lump of modeling clay**
- **plastic wrap**

Do not let anyone use the straw you breathed into.

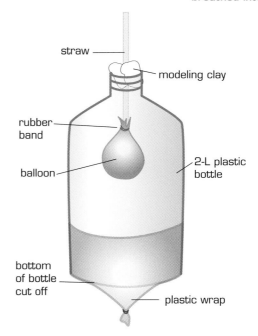

straw

modeling clay

rubber band

balloon

2-L plastic bottle

bottom of bottle cut off

plastic wrap

Connect the Model to the Respiratory System

In a good model, the parts work a lot like the real parts they represent. In your respiratory system model, the parts work a lot like parts of the respiratory system. The different parts of the model you made in class are similar to the real structures or organs they are modeling.

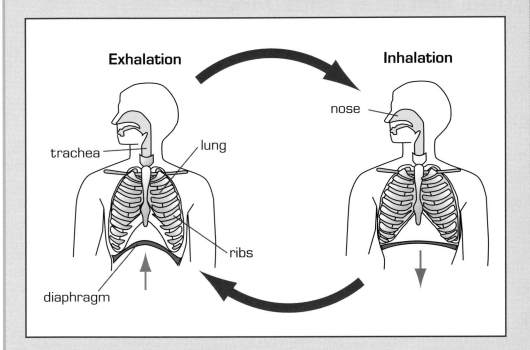

diaphragm:
a band of muscle that regulates pressure in the chest cavity.

The **diaphragm** is a band of muscle. It is involved in the breathing process. Look at the labeled pictures of how the respiratory system works. Find the diaphragm. When the diaphragm contracts, it flattens (goes down). The volume of the chest expands. Air moves into the lungs. Try it. Think about what is happening. Remember back to your respiratory-system model.

This process is similar to what happened when you pulled on the balloon in your model of the respiratory system. When you pulled the balloon down, the bottle cavity became bigger and air entered the bottle.

The balloon in the bottle represented the lungs. A balloon is a good model because air can be used to fill the balloon similarly to the way air can fill the lungs. The balloon can hold air and expand just like your lungs do.

You do not usually concentrate so much on breathing. However, if you try, you can control how you breathe. Singers, athletes, and people who

play certain kinds of instruments learn how to control their breathing to help them perform better.

The diaphragm, lungs, trachea, and nose function together as a system called the respiratory system. The same process happens every time you breathe.

Producing a pleasing sound on any brass instrument depends on the proper use and control of breathing.

Reflect

With your partner, look at the diagram you followed to build your model. Answer the following questions. Be prepared to share your answers with the class.

1. For each part of the model,
   a) identify the part of the respiratory system it represents (include nose, trachea, lungs, and diaphragm).
   b) discuss why the materials used are appropriate.

2. In what ways is the model you built a good model of the respiratory system?

3. In what ways is the model different from your respiratory system?

4. Now, use the model to help you make predictions about staying healthy. Using the model, predict how bacteria or viruses might enter the respiratory system to make you sick.

Beyond the Balloon Model: Diseases That Impact Lung Function

Many different communicable diseases affect the lungs. Bacteria cause some of the diseases. Viruses cause others. Many bacteria and viruses are in the air. Because you breathe air, it is easy for them to enter your body. There is not much you can do to stop breathing in air. Sometimes though, you get microbes on your hands and pass them to your nose or mouth by putting your hands near those places. The bacteria take the free ride to your nose or mouth and enter your body. Sometimes this can make you sick.

Some germs like flu viruses live in wet and warm places like your respiratory system. The first place germs that affect your respiratory system can enter is your nose. If they invade your nose, you can get a runny, stuffy nose and sneeze. If they get further into your respiratory system, the germs can also affect your lungs. You can get a cough, a fever, and chest pain.

mucus: sticky, wet material in your nose and other organs.

Covering your nose and mouth with a tissue when sneezing can help prevent the spread of germs.

Parts of your respiratory system also help you stay healthy. Many microbes that enter your body do not make you sick, because the **mucus** (sticky, wet material) found in your nostrils and throughout your respiratory system protects you by trapping some harmful particles from the air.

Not all respiratory diseases are caused by infectious agents. Many lung diseases are not communicable. When someone talks about having a "lung disease," they may be referring to a disease caused by something other than an infectious agent. Some lung diseases are caused by things in the environment. People who smoke are more likely to get lung cancer and other lung diseases than those who do not smoke. A disease called black lung disease was once common among miners, who breathed a lot of dirty air in mines.

Stop and Think

1. What is one disease you have learned about so far that impacts the respiratory system and is caused by bacteria?

2. What is one disease that impacts the respiratory system and is caused by a virus?

3. Go back to the *Communicable-Disease Information Table* you made in *Learning Set 2* and fill in parts of it that you know more about now.

Update the *Project Board*

Your respiratory system investigations may have helped you answer some of the questions on the *Project Board*. The question that guides this *Learning Set* is *What happens to you when you get sick?* You now know about the respiratory system and how it is necessary for breathing. However, bacteria and viruses can also enter your body through this system. What information about the respiratory system helps you answer some of your questions about disease and how disease spreads?

Add what you have learned to the *Project Board*. Be sure to include the evidence for your statements in the *What is our evidence?* column. Look back at what you wrote for the *What do we think we know?* column from the beginning of the Unit. What information from this section of the Unit can you use to support what you thought you knew before? How did the information about the respiratory system change what you thought you knew? Add this information to the *What are we learning?* column. If you have new questions you had not thought about before, add them to the *What do we need to investigate?* column.

How can you prevent good friends from getting sick?				
What do we think we know?	**What do we need to investigate?**	**What are we learning?**	**What is our evidence?**	**What does it mean for the challenge or question?**

What's the Point?

The respiratory system delivers oxygen to the body. It also removes carbon dioxide produced by the body processes. You breathe in air through your nostrils, and the air moves through your trachea (windpipe) to your lungs. Your diaphragm, a band of muscle right below your ribs, works to move the air from your nostrils to your lungs. When you breathe, the diaphragm contracts or flattens, moving downward. The volume of the chest increases.

Because the chest has more space for your lungs to expand, air moves into the lungs. You saw this happen in the model you built. When you pulled on the membrane at the bottom of your model, you increased the volume of the bottle. Air flowed into the bottle through the straw and inflated the balloon. Bacteria and viruses can get into your body through your respiratory system. You might breathe them in with the air you breathe, or you might give them a little help by placing the microbes near your nose, mouth, or eyes. The respiratory system is warm and wet. Many different microbes that make you sick thrive in warm and wet conditions. When those microbes reproduce, they can make you ill.

Several of the diseases you have read about affect your respiratory system. Pneumonia is one of those diseases. Pneumonia can be spread from person to person and is caused by both bacteria and viruses. When you get pneumonia, you have a high fever, chills, coughs, and difficulty breathing. You may even have pain in your chest.

3.5 Investigate

How Fast Does Your Heart Beat?

The respiratory system moves air into your body. In the lungs, the carbon dioxide made by your cells is exchanged for oxygen from the air. But how does the oxygen you need to live move from your lungs to the rest of your body? The oxygen your lungs extract from air is carried around your body by your blood. The **circulatory system** moves blood around your body so all the parts of your body can get oxygen. It also carries nutrients (food) and other chemicals to the cells of the body. It carries waste away from the cells.

The **heart** is the organ responsible for pumping blood through the body. The rate at which your heart beats depends on many factors. In this investigation, you are going to look at how exercise affects your heart rate and think about why.

Procedure

1. The circulatory system moves blood with oxygen all around your body. Think about how the oxygen might get from your lungs, where you breathe it into your body, to your big toe. Using an outline of the body, draw the path you think blood with oxygen moves through you to get to your big toe.

2. Use your hands to feel your heartbeat. Try putting your hand on your chest. Can you feel your heart there?

3. Take your index (pointing) and middle fingers, and press them to the side of the middle of your neck and just under your jaw. Move your finger around on your neck until you can find your pulse. This is just like feeling your heartbeat.

> **circulatory system:** an organ system that carries nutrients and other chemicals to the cells of the body and carries away waste; includes the heart, arteries and veins.
>
> **heart:** the organ responsible for pumping blood through the body.

Materials

- stopwatch
- **Heart Rate Data page**
- blank page

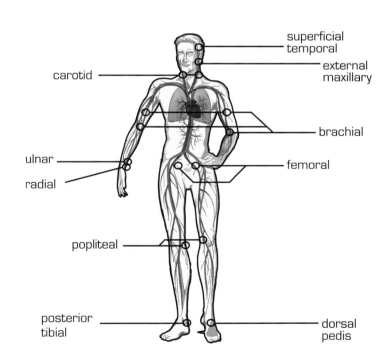

The pulse can be measured anywhere an artery passes close to the skin. The diagram shows several common pulse points.

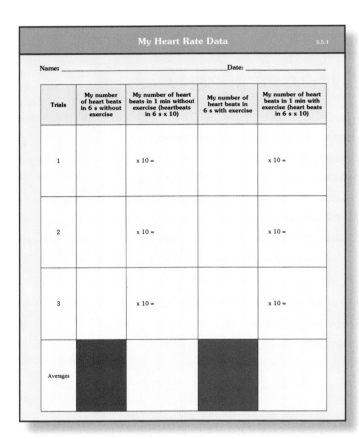

Trials	My number of heart beats in 6 s without exercise	My number of heart beats in 1 min without exercise (heartbeats in 6 s x 10)	My number of heart beats in 6 s with exercise	My number of heart beats in 1 min with exercise (heart beats in 6 s x 10)
1		x 10 =		x 10 =
2		x 10 =		x 10 =
3		x 10 =		x 10 =
Averages				

My Heart Rate Data   3.5.1

Name: _____ Date: _____

4. Work with a partner. Let your partner use the stopwatch while you count your heartbeats. Start the stopwatch. Count your heartbeats until your partner tells you that six seconds is up. Do this a total of three times. Record your data in a table similar to the one below.

5. Repeat, switching roles.

6. Stand near your desk. Make sure you are a safe distance from everyone else. Jog in place for 30 seconds.

7. Sit down and begin counting your heartbeats until your partner tells you that 6 seconds have passed. Begin counting your second trial immediately after the first and your third right after that. Record all three trials on a second table. Then repeat, switching roles.

Analyze Your Data

1. People's hearts beat about 70 to 100 times per minute. Does your resting heart rate fall within that range? If it does not, why do you think it is different? Does your partner's heart rate fall within that range? If not, why do you think it did not?

2. Was your active heart rate higher or lower than your resting heart rate? What about your partner's heart rate? Why do you think the active heart rate is different from a resting one in this way?

Communicate Your Results

Share your pair's results with the class. Create a class table that includes the average resting heart rate for each student and the average active heart rate for each student. The table should look like the one below. You might use a similar table to organize all the data for yourself. Then average the numbers in each column.

Analyze Your Data

Now that you see everyone's data, you can compare your own information to the class information.

1. How did your resting beats per minute compare to that of others in the class?

2. How did your active beats per minute compare with those of the rest of the class?

3. What was the range of resting heartbeats per minute in your class? What was the range of active heartbeats per minute?

4. What was the average resting heartbeats per minute in your class? What was the average active heartbeats per minute?

5. Did the average resting heart rate fall within the range of 85–95 beats per minute? How reliable do you think the data are? Why?

6. You can hold your breath. For a certain amount of time, you can decide not to breathe. You cannot, however, control your heart rate. You cannot tell your heart when to beat or stop it from beating. You can make it go faster, however, and you can make it go slower when it is beating faster. How would you do that?

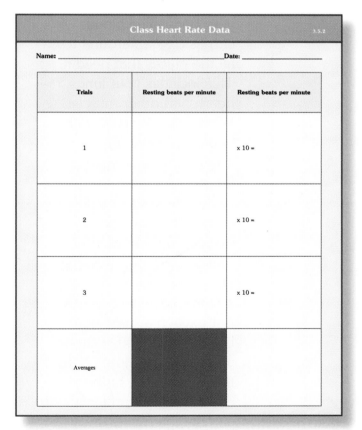

| Class Heart Rate Data | | | 3.5.2 |

Name: _____ Date: _____

Trials	Resting beats per minute	Resting beats per minute
1		x 10 =
2		x 10 =
3		x 10 =
Averages		

Reflect

Your question for this section was *How fast does your heart beat?* Answer the following questions. Be prepared to discuss your answers with the class.

1. The resting heart rates were probably lower than the active heart rates. Why do you think that is? Why would your heart need to beat more often when you are exercising than when you are resting?

2. What was the high number for the resting heart rate? What was the low number? What was the class average number for these rates? Which is the best number to use when you want to know about the population— the high, the low, or the average? Why did you choose that number?

What's the Point?

When you exercise, your heart beats faster and moves more blood around your body. The blood carries oxygen around to all the places that need it. This includes your lungs, heart, and other muscles. When you exercise, your body works harder, so it needs more oxygen. If you don't get more oxygen to your muscles, you will not be able to exercise anymore. Your muscles will stop working. However, your circulatory system moves more than just oxygen around your body. In the next sections, you will learn more about the importance of the circulatory system in getting you sick and keeping you from getting sick.

When you exercise, your heart rate increases to adapt to the body's need for oxygen.

3.6 Investigate

How Does the Circulatory System Work?

In class, you took your pulse to measure how fast your heart was pumping blood. When you took your resting pulse, your heart was working at its normal rate. When you took your active pulse, your heart had been working much harder.

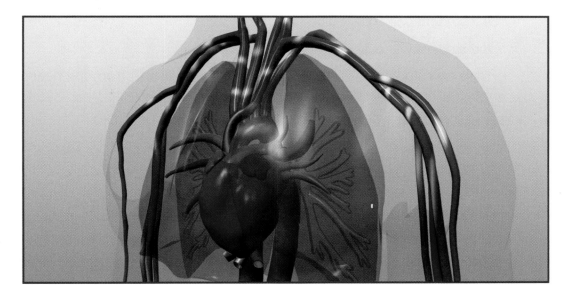

This model of the heart and lungs can help you visualize (see) the parts of the circulatory system and respiratory system that work together to move blood, oxygen, **nutrients** *(the useable substances in food), and wastes through your body.*

nutrients: the useable substances in food.

Your heart is one organ in your circulatory system. It is responsible for pumping blood through your body. One of its functions is to get oxygen to the body cells and carry carbon dioxide away. However, no organ system works on its own. The lungs are part of the respiratory system. They bring oxygen into the body and remove carbon dioxide from the body. The two systems work together to make your body run smoothly.

It is difficult to see what is going on in your body. That is why models are important. You have used models before to see how germs spread and how diseases are transmitted. Now you will model the circulatory system and take part in a simulation to see the path that blood takes around your body. You will see parts of your circulatory system and parts of your respiratory system working together. You need both of these systems to move oxygen and carbon dioxide around your body.

Build and Run Your Model for Getting Oxygen to Your Big Toe

1. Arrange yourselves as follows:

- **Heart (two students)**

    One student will be the left side of the heart (Heart 1).

    One student will be the right side of the heart (Heart 2).

    These two students will stand in the middle of the room.

- **Lungs (two students)**

    One student will collect blue cards (Lung 1).

    One student will pass out red cards (Lung 2).

    These two students will stand about 1.5 m (about 5 ft) from the heart, as shown in the diagram.

- **Big toe (two students)**

    One student will collect red cards (Toe 1).

    One student will pass out blue cards (Toe 2).

Materials

- **100 red cards marked "oxygen"**

- **100 blue cards marked "carbon dioxide"**

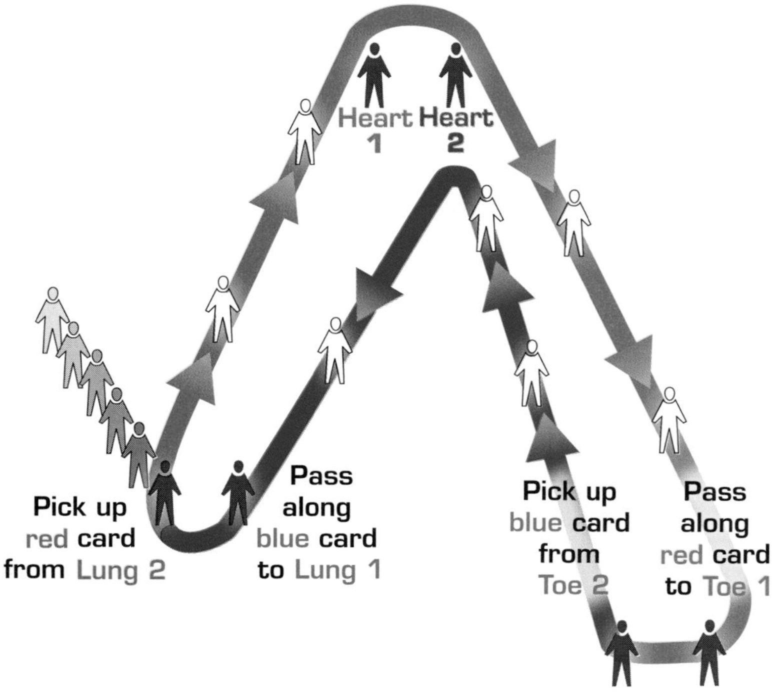

These two students will stand about 5 m (about 15 ft) at an angle from the lungs.

- **Red blood cells (all other students)**

    All other students are red blood cells.

    These students will line up behind the lungs.

    They will receive a red card from Lung 2.

    They will pass a red card to Toe 1.

    They will receive a blue card from Toe 2.

    They will pass a blue card to Lung 1.

Heart Heart
1   2

Pick up
red **card**
from Lung 2

Pass
along
blue **card**
to Lung 1

Pick up
blue **card**
from
Toe 2

Pass
along
red **card**
to Toe 1

GF 98

2. When everyone is in place, the heart students will begin to chant "pump, pump."

3. Each red blood cell will begin by receiving a red (oxygen) card from Lung 2.

4. They will walk toward Heart 1. Heart 1 will redirect them toward the big toe.

5. Each red blood cell will give their red (oxygen) card to Toe 1.

6. Each red blood cell will then receive a blue (carbon dioxide) card from Toe 2. The red blood cell will then go back to the heart.

7. Each red blood cell will pass Heart 2 and clap his or her hands. The "blood" is being pumped back to the lungs.

8. At Lung 1, pass off the blue card (carbon dioxide). Take a new red card (oxygen) from Lung 2 and continue the process.

9. After you have practiced the procedure, try to complete one cycle in 15 s. Your teacher will keep track of the time.

10. Answer the following questions:

   a) The time it takes for your blood to complete one round is 15 s. How close were you able to come to this time?

   b) Your active heart rate is three times faster than this. In what time would you have to complete one round to simulate this heart rate?

11. Try to simulate the blood flow when you are exercising.

blood vessels: tube-like parts of the circulatory system that transport blood through the body.

The main job of the red blood cells is to carry oxygen through the body. Red blood cells look like disks, as shown in the drawing.

Connect the Model to the Circulatory System

In your simulation, the red blood cells that are carrying the oxygen cards move through imaginary paths to get to the big toe. In your body, these paths are called **blood vessels**. At the big toe, they trade in their oxygen cards for carbon dioxide cards. Why does that happen? The blood that travels from the lungs through the heart brings oxygen and other nutrients to your body tissues. On the return trip, the blood brings

GOOD FRIENDS and GERMS

vein: a blood vessel that carries blood back to the heart.

artery: a blood vessel that carries blood away from the heart.

heart chamber: one of four parts–right atrium, right ventricle, left atrium, and left ventricle–of a heart.

atria (singular, atrium): heart chambers that receive blood from the veins.

ventricles: muscular heart chambers that pump blood through the arteries.

waste products, like carbon dioxide, out of the lungs and through the heart. When you ran your simulation, the red blood cells carried oxygen cards to the toe tissue, traded the oxygen cards for carbon dioxide cards, and carried the carbon dioxide back to the lungs, where it could leave the body.

The science model used in this investigation was just like all other models. It has parts that represent the real thing well and other parts that do not represent the real thing well. For example, you tried to get the time it took red blood cells to go from the lungs and back again to be the same as the time it takes blood to flow through the body. This is one way your model is like your body. One way that your model does not mirror your body is that the heart in your model has only two parts. A real heart has four parts to do the job.

Beyond the Model: The Circulatory System
What Makes the Blood Flow Through Your Body?

At the center of the cirulatory system is the heart, which pumps blood so it can flow around through the body. Blood vessels called **veins** and **arteries** carry the blood around your body. Blood leaves your heart and is pumped to your lungs. In your lungs, the blood picks up oxygen. The blood, rich in oxygen, returns to your heart. From there, it is pumped to the rest of your body.

The heart has four parts, or **heart chambers**. The two upper parts of your heart are called **atria** (singular, atrium). They receive the blood from your body (your big toe in your model) or your lungs. The lower chambers are called the **ventricles**. These are the pumping parts of your heart. The muscles in the ventricles are very strong so they can pump blood to distant tissues of your body.

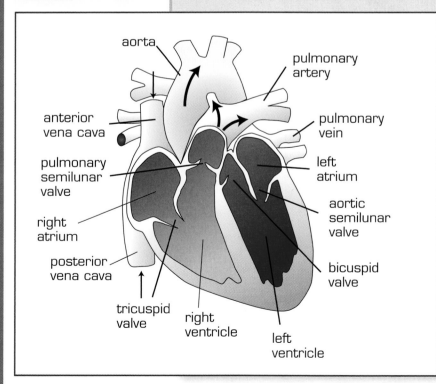

aorta

pulmonary artery

anterior vena cava

pulmonary vein

pulmonary semilunar valve

left atrium

right atrium

aortic semilunar valve

posterior vena cava

bicuspid valve

tricuspid valve

right ventricle

left ventricle

What Paths Does Your Blood Take Through Your Body?

The model is also different than your circulatory system in another way. In the model, you walked or ran by the shortest route from the lungs to the big toe. In your body, your blood does not choose its route. It moves through a network of veins and arteries that direct it to all the places it has to go.

However, there is a similarity in the path that the blood takes in your body and in your model. In your model, the red blood cells moved in one direction only. None went against the flow. The blood flow in the body is also only one way.

The arteries carry blood away from the heart. Every time your heart beats, blood is pushed from the heart into the arteries. The arteries stretch to accommodate this surge. The **pulse** you felt in your neck is caused by the change in diameter of the arteries near the surface of your body as the blood surges forward.

Blood from the heart passes into large arteries. From there it passes into smaller and smaller arteries. These eventually lead to extremely tiny blood vessels. These very tiny blood vessels connect the arteries to the veins. The veins are very small at first. However, they become progressively larger as the blood flows back to the heart. The blood returns to the heart through large veins.

pulse: the surge of blood in an artery as the blood is pumped by the heart.

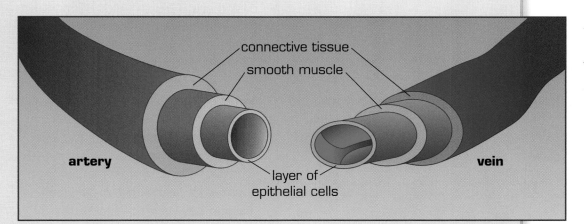

artery

vein

connective tissue

smooth muscle

layer of epithelial cells

Arteries carry blood away from the heart, while veins carry blood to the heart.

How Does the Circulatory System Contribute to Sickness?

There is one more important way the model is different from your circulatory system. In the model, the blood cells carried oxygen to parts of your body and carried carbon dioxide back to your heart and lungs. In your body, the blood also carries many other things. It carries nutrients (vitamins and minerals) around your body that the cells in your body need to function. You'll learn more about how those nutrients get into the blood when you learn about the digestive system. Your blood can also carry germs (bacteria and viruses) around your body.

Do you remember going to the doctor to get a tetanus shot? Tetanus, also called lockjaw, is a bacterial disease that infects your blood. It causes your jaw to lock up so you can't move it. If you fall down or cut yourself and bleed, the microbes in the environment can get into your bloodstream through your wound. You get tetanus shots to protect you from tetanus bacteria that can get into your body if you cut yourself on something rusty.

What Color Is Blood?

Look at your arm where your elbow bends. The skin there is thin, so you can see the veins and arteries. If you look carefully, you will see that these veins and arteries seem to be different colors. They may be deep red, lighter red, or almost blue. Some people may have told you that the tubes look different because the blood inside them is blue. Some people think that the blood is blue because it does not have oxygen in it. But that is not true. Blood is always red.

When you donate blood, it is taken from a vein in your arm. The blood that flows through the needle and into the collection bag is very dark red.

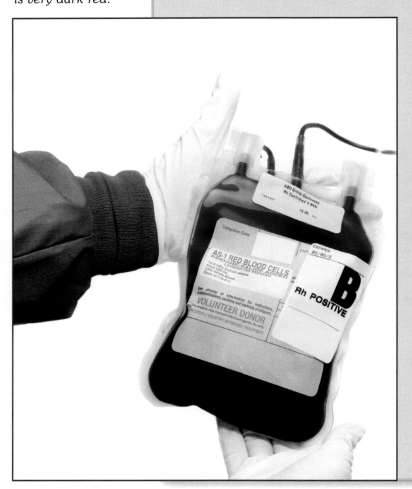

Veins are a bluish color because you are looking at them through your skin. The layer of skin on top of veins and arteries affects the color you see. The blood in your arteries has just been through your lungs and is full of oxygen. This makes it bright red. You see that as red through your skin. The blood in your veins is dark red instead of bright red. It has less oxygen because the blood has left its oxygen in other parts of your body. As it is pumped around your body, your blood loses more and more of its oxygen. As it loses oxygen, it becomes darker. When the blood gets very dark, it appears bluish through your skin.

Reflect

You now know more about the circulatory system than you did when you simulated it earlier. If you could run another circulatory system simulation, what would you change to more accurately model the circulatory system? Be prepared to share your ideas with the class.

Revise your Model

As a class, decide how to better model and simulate the circulatory system.

Update the *Project Board*

Your circulatory system investigations helped you better understand how the system works. The question that guides this section of the Unit is *What happens to you when you get sick?* You now know about the circulatory system and how it is necessary for moving nutrients, oxygen, and waste through your body. However, it can also move bacteria and viruses around your body. What information about the circulatory system helps you answer some of your questions about disease and how disease spreads?

Add what you have learned to the *Project Board*. Review all the information you have already added to the board. Look back at what you wrote in the *What do we think we know?* column in the beginning of the Unit. What information from this section of the Unit can you use to support what you thought you knew? How did the information about the circulatory system change what you thought you knew? Add this information to the *What are we learning?* column. Be sure to include evidence for your statements in the *What is our evidence?* column.

What's the Point?

The circulatory system works with the respiratory system to move oxygen, nutrients, and waste around your body. Oxygen, nutrients, and waste are carried by the blood. The heart acts as a pump. It receives blood from the lungs, which contains red blood cells full of oxygen. The heart pumps the red blood cells through the arteries, which carry them to all the other cells in your body. The blood returns through the veins. This time, the red blood cells carry carbon dioxide produced by the body cells. When blood with carbon dioxide reaches the heart, the heart pumps it back to the lungs. The lungs remove the carbon dioxide from the red blood cells and fill them again with oxygen.

The circulatory system is a very complicated system in your body. It is very important in making sure your body maintains its health. The movement of oxygen and nutrients throughout your circulatory system is critical to breathing and having healthy muscles. But the circulatory system does not make a choice about what travels around your body in your blood. Sometimes, bacteria and viruses move through your circulatory system. The way your circulatory system works is usually a great benefit to you. Sometimes, however, it can make you feel sicker as it spreads germs around your body.

3.7 Investigate

How Does the Digestive System React to Illness and Help Spread Illness?

The respiratory and circulatory systems work together to move oxygen around your body. You read in the last section that diseases can move into your body and then move around your body through your respiratory and circulatory systems. Sometimes diseases that enter through your mouth and nose end up making your digestive system sick.

What's in Food?

If your digestive system is not working properly, it cannot digest or break down food. Foods contain a variety of substances that are chemically broken down by enzymes into small components. Your body cells can then use these small components. **Carbohydrates** and **fats** are two substances found in food. When broken down, they give your body the energy it needs. The energy in food is measured in **Calories**. A Calorie (C) is the amount of energy in foods. There are 1000 **calories** (c) in 1 Calorie or kilocalorie. (A calorie is the amount of energy needed to raise 1 gram of water 1°C. The more Calories a food has, the more energy it contains. Your body needs a certain number of Calories each day to meet your body's energy needs. **Vitamins**, needed in small amounts, help your body in many ways, such as keeping you from developing certain diseases. **Minerals** such as calcium are needed for strong bones and teeth and for healthy red blood cells. **Proteins** build muscle and body tissue, and water is needed to carry out all the cell functions.

Using common materials, you are going to build a scale model of the digestive system. Remember that one role models can play is to help you see things that are difficult to see. Your digestive system is difficult to see because it is inside you.

carbohydrate: a type of substance found in food that gives the body the energy it needs.

fat: a type of substance found in food that gives the body the energy it needs.

Calorie: the amount of energy in foods. One Calorie is the same as 1 kilocalorie or 1000 calories.

calorie: the amount of energy needed to raise the temperature of one gram of water by 1°C.

vitamin: a type of substance found in food that is needed in small amounts. It helps the body in many ways, such as keeping it from developing certain diseases.

mineral: a type of substance found in food that is needed in small amounts. The mineral calcium is needed for strong bones and teeth and the mineral iron is needed for healthy red blood cells.

protein: a type of substance found in food that builds muscle and body tissue.

Procedure

Someone will read the story below. As you listen to the story, you will use the materials listed to build a model of the digestive system. Consider how the materials mirror the actual parts of your digestive system.

Imagine sitting with your friends at a pizza parlor. You have just ordered your favorite pizza and you are waiting patiently. You notice that your mouth has started to water. You are thinking about the pizza, and your mouth makes more saliva. Before you even begin eating, you have started the digestion process. An enzyme in saliva chemically breaks down carbohydrates into smaller parts (sugars and starches).

Materials

- 1 round balloon
- 8.5 m (about 25 ft) of long balloons or rubber tube
- construction paper
- mouth and Teeth page
- tape
- poster board

esophagus: the tube that carries food from the mouth to the stomach.

1. Take the *Mouth and Teeth* page and write the word "saliva" in the mouth.

When the pizza comes to the table, you dive into eating it. You take a big bite with your front teeth and tear off a piece. Where do you chew the pizza? You probably chew with your back teeth. Molars, your back teeth, are excellent for breaking the food into much smaller pieces and mashing it. Along with the saliva in your mouth, your teeth make it possible for you to swallow the food and begin the digestion process.

2. On the *Mouth and Teeth* page, write "tear" on the teeth in the mouth that help you tear food. Label the teeth in the mouth that help you chew food into smaller pieces with the word "chew." Tape the *Mouth and Teeth* page to the poster board.

*The pizza pieces, now much smaller, travel down your **esophagus**. The esophagus is the tube that leads to your stomach. This tube pushes the food into your stomach. It takes about 10 seconds for food to get pushed down this tube.*

3. Take one long balloon and attach it to the picture of the head at the back of the mouth.

As the pizza enters your stomach, at the end of your esophagus, your stomach begins to churn. The churning is one way your stomach makes the food you have swallowed even smaller. Your stomach has special cells that make enzymes. These enzymes are sent into your stomach when you eat food. They break

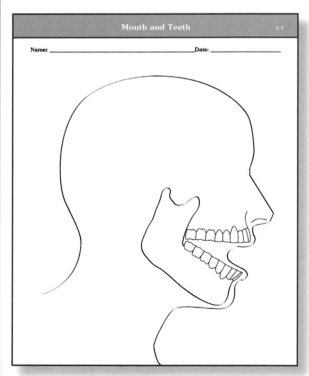

Mouth and Teeth

Name: _____ Date: _____

the proteins in the food into smaller substances. Your food gets mashed by the churning of your stomach and digested by the enzymes that are mixed with it.

Your stomach works on the food for a while. Your stomach stretches to fit more and more food. It is similar to a balloon. When it needs to hold food, it can stretch. When it doesn't need to hold food, it is much smaller. A small opening at the bottom of the stomach remains closed as your food is digested. When the food has been digested enough, the small hole that was closed opens, so the food can leave. The digested food moves into your small intestine.

4. Use the round balloon to represent your stomach. Decide if the stomach is full or empty before you attach it. If you want a full stomach, blow up the balloon and tie a knot to keep in the air. Then attach the balloon to the end of the esophagus with tape.

villi (singular, villus): tiny, finger-like structures that protrude from the inside surface of the small intestine.

Your small intestine is the next organ in the digestive system. Food moves from your stomach into your small intestine, where more enzymes break down carbohydrates and proteins into smaller substances. Fats are also broken down into smaller substances in the small intestine. The goal of all the parts of your digestive system is to make the food as small as possible so your body will be able to use it.

*The small intestine is very long. It can be up to 7 m (about 23 ft) long, but it is only 2.5 cm (about 1 in.) in diameter. This tube squeezes the food along. Tiny finger-like structures, called **villi**, cover the inside wall of the small intestine. These structures allow the digested parts of carbohydrates, proteins, and fats to be carried out into the bloodstream. Other nutrients like vitamins and minerals also pass through the villi. The blood then carries these nutrients to all parts of the body.*

GOOD FRIENDS and GERMS

5. Take 7 m (23 ft) of balloons and tape them to the stomach, near the bottom. The balloons represent the small intestine. Lay the intestines on your poster and carefully arrange the balloons (intestines) so they do not get knotted up.

Your body does not use all the bits of food. Some of the food moves on to the large intestine. This is a 1.5 m long (about 5 ft) tube that is attached to the end of your small intestine.

*The small bits of food and water that enter the large intestine are called waste. When the waste enters your large intestine, it is full of water and other liquids. Your large intestine removes the water and other liquids from the waste. What is left is called **feces**. The feces are pushed out of your body through your **anus**. These are the wastes that you flush down the toilet.*

*Both your small and large intestines are organs in your digestive system. They are filled with bacteria that help with the digestion process. These bacteria are called **enteric bacteria**. Enteric means "in the intestines." Almost all of the bacteria in your intestines are good for you. They help you digest food. They also allow your intestines to absorb different vitamins and minerals.*

6. Use 1.5 m (5 ft) of balloons to represent your large intestine. Attach them to the end of the small intestine. Again, arrange them on the poster so that they are not twisted or knotted.

7. Label each organ in your model. Make sure you know which parts are the esophagus, the stomach, the small intestine, and the large intestine.

feces: waste that is produced in the digestive system.

anus: the bottom opening of the digestive tract.

enteric bacteria: bacteria that live in the digestive tract.

Did You Know?

One "famous" bacterium that lives in your intestines is called *E. coli*. *Escherichia coli* is named after the bacteriologist Theodor Escherich, who first discovered it. You read earlier that *E. coli* can cause gastrointestinal illness. However, many of the *E. coli* bacteria that live in your intestine will not make you sick.

E.coli bacteria on a lettuce leaf.

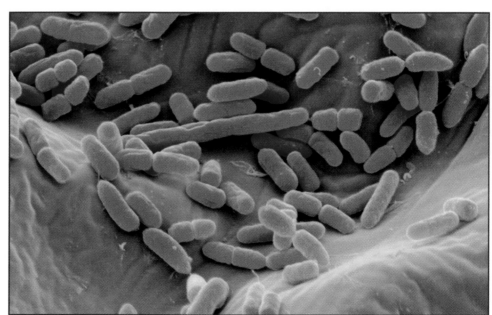

Reflect

In this activity, you built a model of the digestive system. This model helped you see what you cannot normally see.

1. You labeled the parts of the digestive system with their names. Now record the function (job) of each part on your poster.

2. In the model, some of the parts were good representations of the organs of your digestive system. Like all models, however, some of the parts do not represent the system well.

   a) Choose one part of this model that does a good job of representing an organ in the digestive system. Describe why you think so.

   b) Choose one part of this model that does not do as good a job of representing an organ in the digestive system. Describe why you think so.

Beyond the Model: Your Digestive System and Circulatory System Work Together

Every cell of every system in your body relies on the digestive system for nutrients. You learned that the circulatory system works with the respiratory system. The circulatory system also works with the digestive system. Your blood absorbs the food that is broken down by the digestive system. Your blood vessels then transport the food to all the cells of the body.

Other systems in your body also work with the digestive system. Muscles and bones (the skeletal and muscular systems) are used to take in food. For example, muscles in your face help you tear and chew food. The nervous system and endocrine system, which are other systems in your body, regulate how the digestive system functions.

How do people get digestive-system diseases? How do they spread them to others? Sometimes, digestive diseases can come from your hands. You may touch something with harmful bacteria or a virus on it. If you then hold your fingers near your face or put them in your mouth, the microbes can enter your digestive system through your mouth.

Symptoms of digestive-system illnesses are vomiting and diarrhea. These are ways the body reacts to rid your body of these microbes so you will get well. When you are sick with a digestive disease, the waste products from your body can have in them the bacteria or viruses that made you sick. Like *E. coli*, these bacteria live in your digestive

GOOD FRIENDS and GERMS

system. Every year, many people around the world become sick from communicable diseases of the digestive system. Managing the spread of these types of diseases is an important health issue. They are removed in your waste products when you use the bathroom. These bacteria or viruses can get on your hands. If you do not wash your hands, you do not get rid of the bacteria or viruses. They can then be spread directly to other people or by getting on surfaces. (Remember the glow-powder investigation?) Suppose someone who didn't wash his or her hands after using the toilet touches a doorknob. Then if you touch it, you could get the bacteria or virus on your hands. If you put your fingers near your face or in your mouth, you may get the disease.

If not handled properly, chicken is one type of meat that can make people sick. Wash your hands thoroughly before and after handling raw chicken.

Every year, many people around the world become sick from communicable diseases of the digestive system. Managing the spread of these types of diseases is an important health issue.

Some types of digestive-system communicable diseases start in food. When harmful microbes, often bacteria, are in food you eat, they enter your body. Bacteria that need warmth and wetness reproduce in your body. When there are a lot of them, they can make you sick. That's why it is important to keep meat and milk products and some other foods cold in the refrigerator.

Use a different plate for cooked meat than you used for raw meat.

It is also important to cook meat well. Meat can have bacteria on it from when it was processed. Cooking the meat well kills the bacteria. But then the meat can get more bacteria on it from the air. Suppose you leave meat that has bacteria on it out of the refrigerator for a long time. Those bacteria can live and rapidly reproduce in the warm air. If you eat the meat, the bacteria will make you sick. A refrigerator is cold enough so bacteria reproduce very slowly. Food kept in the refrigerator is less likely to have large amounts of harmful bacteria in it.

Some digestive-system diseases are not communicable. One disease of the digestive system that is not communicable is gallstones, a gallbladder disease. The gallbladder is a small, pear-shaped organ on your right side, under your liver. It stores a chemical called bile. Bile is made in your liver. Your digestive system uses bile to help digest fats. Sometimes substances in the gallbladder form gallstones. Gallstones can cause severe pain, nausea, and vomiting. But gallstones are not communicable; you cannot develop gallstones from being with someone who has them.

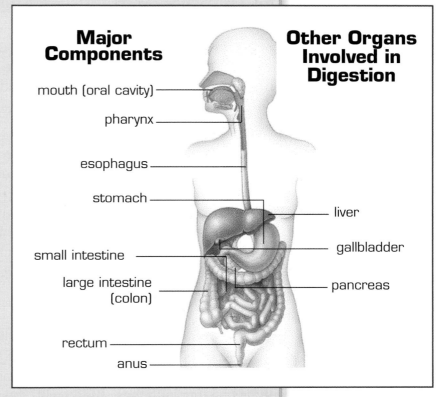

Major Components

- mouth (oral cavity)
- pharynx
- esophagus
- stomach
- small intestine
- large intestine (colon)
- rectum
- anus

Other Organs Involved in Digestion

- liver
- gallbladder
- pancreas

The digestive system includes many organs. They work together to help your body use the food you eat and help you grow and stay healthy. If one organ does not function properly, the whole digestive system cannot function properly.

Stop and Think

1. Go back to the *Communicable-Disease Information Table* you made in *Learning Set 2*. Fill in parts of it that you now know more about.

2. What are some good ways to reduce the chances of spreading digestive-system communicable diseases? Follow the path of disease-causing

bacteria through the digestive system and decide what you would suggest to someone to prevent the spread of disease.

3. Choose one of the organs and describe what you think would happen to the digestive system if the organ stopped working correctly.

Update the *Project Board*

Your digestive system model building and reading may help you add more to the *Project Board*. The question for this section is *What happens to you when you get sick?* You now know about the digestive system and how it reacts to bacteria and viruses. You even have some ideas about how the digestive system moves bacteria and viruses out of your body and increases the chances that you could spread illness to someone else. What information about the digestive system do you think you should add to the *Project Board*?

Add what you have learned to the *Project Board*. Be sure to include evidence for your statements in the *What is our evidence?* column. Look back at what you wrote for the *What do we think we know?* column from the beginning of the Unit. What information from this section of the Unit can you use to support what you thought you knew before? How did the information about the digestive system change what you thought you knew before? Add this information to the *What are we learning?* column.

What's the Point?

The digestive system is responsible for digesting or breaking down food. Food enters your digestive system through your mouth. The saliva in your mouth contains an enzyme that chemically breaks down carbohydrates into sugars and starches. Along with the saliva in your mouth, your teeth tear and mash food so it possible for you to swallow food and begin the digestion process. Next, smaller pieces of food travel down a tube called the esophagus that connects to the stomach. Once food enters the stomach, it begins to churn the food and release enzymes that break down proteins into smaller substances. The next digestive organ that food travels into is the small intestine where more enzymes break down carbohydrates and proteins into smaller substances. Finger-like structures called villi cover the inside wall of the small intestine. They allow digested parts of carbohydrates, proteins, and fats as well as vitamins and minerals to be carried out into the bloodstream to all parts of the body. Lastly, the undigested parts of food move into the large intestine and are pushed out of the body. The large intestine also removes water from the wastes.

3.8 Read

How Does Your Body Protect Against Disease?

The next body system you will learn about is the **immune system**. This body system's function is to protect against disease. The immune system detects bacteria and viruses in your body and fights them off. Sometimes it can do that before you even feel anything. After you've been sick, some cells in your immune system continue to live for many years and remember the bacteria or virus they fought off earlier. Those cells can give you **immunity** from that same bacteria or virus. Immunity means that you will not become sick by that germ again.

immune system: body system that fights disease.

immunity: a condition of being able to resist an infectious disease.

The immune system is very complicated. In all the other systems you have learned about, the parts are connected to each other physically, and parts of the system move substances to other parts of the system. In the immune system, it is different. The immune system has only some parts that are connected to each other and have the specific job of fighting disease. In addition, it depends on other parts of your body. For example, your skin is very important to protecting you from disease. It acts as a barrier to keep bacteria and viruses from entering your body. If germs can be kept from your immune system, it does not have to work as hard.

You will begin learning about how your body defends against disease by thinking about a time when you were sick.

What Does It Mean to Protect Against Disease?

Think back to a time when you had a cold or the flu. What symptoms did you have? Did you have a runny nose, scratchy throat, stomachache, headache, or fever? Maybe you had other symptoms, as well. Why did your body react the way it did?

Your body reacts the way it does because it is trying to protect you. What good does it do to have stuff running out of your nose? When your nose runs, you know your body is fighting off bacteria, viruses, or other germs. Mucus serves a purpose. It traps germs. When you have a cold and you blow your nose, you remove some of the mucus from your body along with the germs trapped by the mucus.

The body has several defenses against germs. Skin keeps germs from getting into your body. Mucus linings in the breathing passages work to keep germs from getting through your nose and into your respiratory system. White blood cells and other substances attack microbes to kill them after they are in your body. Small nodes and glands all over the body help white blood cells develop and reproduce. Other glands **filter** microbes from your body fluids and trap them or **expel** them from your body. Together, these parts form the body's immune defense system.

The immune system defends the body against **microscopic** living things, like bacteria and viruses, that might make you sick. You know that even in a clean and spotless place, there are many bacteria and viruses. They get onto your skin, into the food you eat, into the drinks you consume, and even into the air you breathe. These bacteria and viruses may also get into your body through a cut or scrape on your skin, through your nasal passages, or through the throat. You get sick when the bacteria and viruses that get into your body have a chance to reproduce and multiply. When that happens, glands in your body work overtime to create more **white blood cells**, or **leukocytes**, to kill the germs and to filter the germs from your body fluids.

How Does Your Body Attack and Kill Germs?

The **lymphatic system** is part of the immune system that collects fluid lost by the blood and returns it to the circulatory system. The organs of the lymphatic system include lymph nodes and lymph vessels.

The lymphatic system also produces several different kinds of white blood cells. These are the cells that hunt down and kill diseases. They also protect you from diseases. There are many types of white blood cells. Each type has a special job in protecting you from disease. Some white blood cells wrap themselves around invading bacteria to trap and suffocate them. Some white blood cells, called killer T-cells, recognize when a cell has been invaded by a virus and punctures holes in those cells. When that happens, the infected cell's cytoplasm leaks out, and the cell dies along with the virus that invaded it. Another kind of white blood cell acts as a vacuum cleaner. They eat the remains of the dead cells.

Some white blood cells, called B-cells, produce **antibodies** that are specialized to the shape of invading microbes. The antibodies then attach themselves to the invaders. They might kill the invader or attach themselves in a way that inactivates the invader (keeps it from doing its job). Some

filter: to separate.

expel: to force out.

microscopic: so small you can see it only through a microscope.

white blood cells (leukocytes): blood cells whose main job is to protect the body from invading microbes.

lymphatic system: body system that collects fluid lost by the blood and returns it to the circulatory system. It also produces some types of white blood cells.

antibodies: substances that help destroy harmful microbes.

antibodies keep watch for the kind of invader they specialize in. Remember Edward Jenner? His experiments with vaccination relied on these memory cells. When they recognize a familiar cell, they send out a signal, and other white blood cells come and kill those invading cells. Because of memory cells, the second time your body encounters some bacteria or virus, your body attacks it more quickly. This is how many vaccines work.

The organs in your lymphatic system produce white blood cells, help them develop and reproduce, and carry them around your body. They also filter or trap bacteria, viruses, and dead cells to remove them from your blood.

Bone Marrow

Your **bone marrow** is the soft material in the center of your bones. All your blood cells, including white blood cells, are produced in bone marrow. The long bones in your body produce most of your blood cells.

Thymus

Your **thymus** is an organ in the immune system. The thymus is located high in your chest, beneath your breastbone and between the two parts of your lungs. It is important in the development of some types of white blood cells. Some of the white blood cells produced by your bone marrow are ready to go to work right after they are created. Others need to mature or develop more complicated structures. The thymus helps the kind of white blood cells called T-cells to develop.

Spleen

Your **spleen** is located under your ribs and in front of your lungs on the left side of your body. This organ has many different functions. It filters your blood to remove bacteria or viruses that might make you sick. B-cells are activated in the spleen. Old red blood cells that cannot do their job anymore are destroyed in the spleen. If somebody has to have their spleen removed, other organs in the immune system take over the spleen's functions. But the spleen carries out those functions better than any other organs can.

Lymph Vessels

Your **lymph vessels** carry the different kinds of **lymphocytes** around your body. Lymphocytes are a type of white blood cell that makes antibodies. Both T-cells and B-cells are lymphocytes. Lymph vessels are very much like blood vessels, and they can be found right next to blood vessels in your body.

bone marrow: a soft tissue in the center of bones where most blood cells are produced.

thymus: an organ that is important in the development of a type of white blood cell.

spleen: filters blood by removing harmful bacteria and viruses, activates B-cells, and destroys old red blood cells.

lymph vessels: structures that carry lymphocytes throughout the body.

lymphocyte: a blood cell that protects the body from invading microbes.

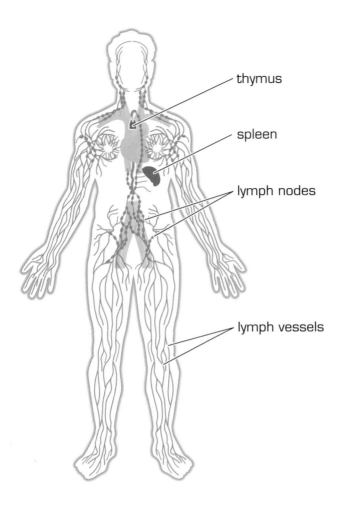

thymus

spleen

lymph nodes

lymph vessels

Lymph Nodes

Lymph nodes are all over your body. They can be found along the pathways of your lymph vessels. They are in your neck, under your arms, and in your groin and chest. Like the spleen, the lymph nodes have many functions. One of their functions is to hold large numbers of lymphocytes for when they are needed to fight infection. When you have an infection, the lymph nodes become warm and provide a warm place for lymphocytes to reproduce actively so there are enough available to fight your infection. When lymphocytes reproduce quickly, they can build up in your lymph nodes. This makes the lymph nodes swell. When you have a cold, the lymph nodes in your neck sometimes swell enough so you can feel lumps on either side of your neck. When you are sick and go to the doctor, he or she may feel each side of your neck to see if the lymph nodes are swollen. The lymph nodes also act as filters, like the spleen.

lymph nodes: structures in the lymphatic system that filter lymph and trap microbes.

barrier: something that obstructs (or blocks).

How Else Does Your Body Defend Against Illness?

If the lymphatic system had to fight every bacteria and virus in the air or on things you touch, it would have too much work to do. Your body helps the lymphatic system in several ways:

- Your body provides **barriers** that keep harmful microbes from getting into your body. Your skin is one of those barriers. The mucus in your nose is another. In fact, you have mucus on the insides of organs in your respiratory system, your digestive system, and other systems. Mucus forms a barrier. A layer of mucus helps keep microbes from getting through to healthy tissue. When some body part with mucus gets irritated, it secretes additional mucus.

- Some of your body systems make substances that break down the cell walls of bacteria. The saliva in your mouth does that, and so do your tears.

- All of your cells have an **inflammatory response**. When a cell is invaded by a microbe, it gives off a chemical signal (called a histamine) that increases blood flow to the area. That area becomes red and heated. A heated area of your body provides a good environment for white cells to reproduce and enough heat to kill some of the invading microbes.

- Some body systems have particular inflammatory responses. For example, when your stomach is inflamed, you vomit. When your intestines are inflamed, you get diarrhea. These inflammatory responses, or **inflammations**, expel harmful microbes from your body.

Stop and Think

1. Sometimes when you are very sick, the doctor will take some blood and count the different kinds of cells in your blood. Why do you think he or she might count your white blood cells?

2. The immune system is very complicated and has many parts. The parts work together to keep you healthy. If the parts of this system did not work your body would not be able to fight off infections. Even the smallest infection could make you very sick. Describe in your own words how the parts of the immune system work together to fight disease.

inflammatory response: the body's reaction to invading bacteria or viruses; includes the release of a chemical signal and increased blood flow.

inflammation: a reaction to invading microorganisms; signs of this include redness, heat, swelling and pain.

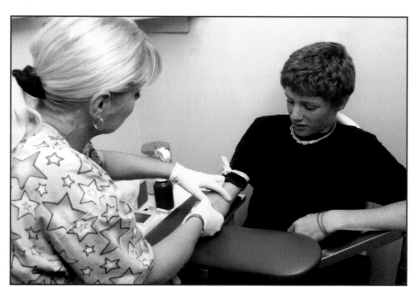

Sometimes when you are sick, a doctor may order a WBC count. This is a blood test to measure the number of white blood cells.

GOOD FRIENDS and GERMS

Communicate Your Understanding

Choose one part of the immune system. Make a poster showing the following three things:

- what this part of the immune system does to protect agaist disease

- what other parts of the immune system depend on it

- what other parts of the immune system it helps

Each group will have a chance to show their poster. As you listen to the presentations, take notes about how the different parts of the immune system work together.

After the presentations, work as a class to figure out how the immune system organs work together to keep you healthy.

Go back to the *Communicable-Disease Information Table* you made in *Learning Set 2* and add the new information.

Update the *Project Board*

The question for this section of the Unit is *How does your body protect against disease?* You now know about the immune system and how it is devoted to maintaining good health. What information about the immune system helps you answer some of your questions about disease and how disease spreads?

Add what you learned from the presentations and reading about the immune system to the *Project Board* in the *What are we learning?* column. Be sure to include evidence for your statements on the *Project Board*.

Look back at what you wrote at the beginning of the Unit in the *What do we think we know?* column. What information from learning about the immune system can you use to support what you thought you knew before? How does what you have read about the immune system change what you thought you knew before? Add this information to the *What are we learning?* column.

What's the Point?

Sometimes bacteria or viruses make your body ill. When you are sick, your immune system fights against these germs. Sometimes your lymph nodes, especially the ones in your neck, swell up. When you are healthy, they are about the size of a pea or grape, and you can barely feel them. During illness, they can be as big as golf balls.

When your lymph nodes swell, it means that your immune system is doing important work. The immune system includes white blood cells that attack germs. If bacteria or viruses get into the body, the immune system tries to find them and get rid of them before they can reproduce. Lymph nodes contain billions of white blood cells that multiply rapidly to fight invading germs.

Your lymphatic system would have too much to do if your body did not also have other defenses against germs. Those defenses include barriers, secretions, and inflammatory reactions.

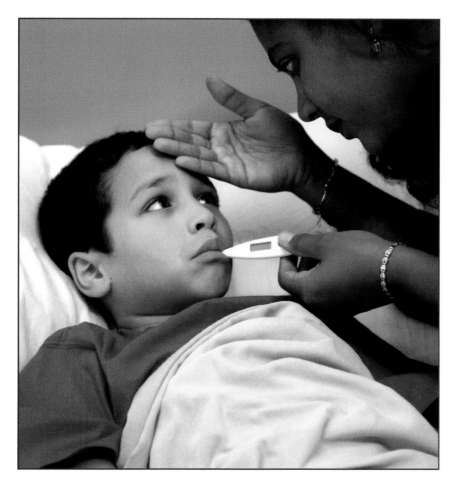

A fever is often a symptom of an infection. Viruses and bacteria have a hard time surviving at temperatures higher than body temperature. A fever is the body's way of fighting an infection. Fever increases blood flow and speeds up the body's defense mechanisms.

Learning Set 3

Back to the Big Question

How can you prevent your good friends from getting sick?

Your body is one large system made up of many smaller systems. You explored the systems within your body that make it possible to breathe; to move oxygen, nutrients, and wastes around; and to digest food. You also learned how your body systems protect you from disease. Each of the systems in your body interacts with other systems. They work together to keep your body healthy.

Beginning with cells in *Learning Set 2*, you began to see the wide variety of systems in your body. Each cell is a small but complicated system. Specialized cells perform specialized functions, and those cells cooperate to form tissues. These tissues are then grouped to form organs. Organs are grouped together to make up organ systems. Each type of cell, tissue, organ, and organ system has a distinct structure and set of functions. Only when your body systems are all working together with each other does your body function in a healthy way.

The systems you learned about, from cells to your whole body, show that the job a system can do is related to the way it is built. Lungs help you breathe because they can move in and out to allow air in and push it out. The parts of your heart, the chambers, each have a special job and are built a particular way to do their jobs.

You will finish this *Learning Set* by exploring the ways different body systems work together with each other and then revising your explanations about how we get sick with communicable diseases and your recommendations for staying healthy.

Each group will get a set of four transparencies and special pens for writing on the transparencies. Each will have the outline of a body. Each person in your group will choose a different body system and draw that body system on their transparency.

The hardest part of doing this is that when you lay all of the transparencies your group draws on top of each other, the organs will need to be in the right places. This means that, as you are doing this activity, you will have to talk to each other about where the different organs go.

Make sure you include all of the organs in each body system. This might be hard, but around the room are posters showing each system, and there are pictures throughout this book.

Communicate Your Ideas

Solution Showcase

Each group will have a chance to show their transparencies to the class and present their solution. When you show your drawings, report to the class about the following four things:

- What did you find difficult about the activity?

- How did you work with your group to make the systems fit the body?

- What organs or systems were most difficult to identify, draw, or locate?

- How does your drawing help you better realize the complexity of how your body works?

Afterward, as a class, discuss the reasons this activity was so challenging. Make a list of the complexities of the human body and how it works that you find most interesting or surprising.

Explain

It is time now to explain how we get sick with communicable diseases using what you have been reading about body systems. Each group in the class will focus on a different body system and develop claims and explanations to answer some of the following questions:

- How does the body system your group is focusing on keep you healthy?

- How do people get sick through the body system your group is focusing on?

Answer whichever questions are applicable to the body system your group is focusing on. As always, begin by developing a claim to answer one of the

Create Your Explanation

Name:_____ Date:_____

Use this page to explain the lesson of your recent investigations.

Write a brief summary of the results from your investigation. You will use this summary to help you write your Explanation.

Claim – a statement of what you understand or a conclusion that you have reached from an investigation or a set of investigations.

Evidence – data collected during investigations and trends in that data.

Science knowledge – knowledge about how things work. You may have learned this through reading, talking to an expert, discussion, or other experiences.

Write your Explanation using the *Claim*, *Evidence* and *Science knowledge*.

questions, and record it on a *Create Your Explanation* page. Then record evidence from the simulations you have been running and science knowledge from your readings to support your claim. Develop a statement that ties together your claim, your evidence, and your science knowledge. Make sure that your statement describes enough about how your body system works so that others will be convinced of your claim. Remember that a good explanation can convince someone else that your claim is trustworthy. If your statement doesn't seem convincing, revise your claim so your evidence and your science knowledge support it.

Recommend

Using your claims and explanations about how your assigned body system keeps us healthy and how we get sick through that body system, develop recommendations for staying healthy. Focus on answering the questions below that are relevant to the body system you are assigned.

- How can I keep from getting sick through the body system I am focusing on?

- How can I prevent friends from getting sick when there is an infection in the body system I am focusing on?

Use a *Create Your Explanation* page for each recommendation. Record your recommendation as a claim, and then record evidence and science knowledge that support it. Develop a statement linking your claim to the evidence and science knowledge that support it. Remember that a good explanation can convince someone else that your recommendation is good. If your statement doesn't seem convincing, revise your recommendation and the evidence and science knowledge that support it so that you can make a more convincing statement.

Communicate

Share your Explanations and Recommendations

Share your group's claim and explanation with the class. Then share the recommendations you have developed about staying healthy. For each claim and recommendation, tell the class what makes your claim accurate based on your evidence and science knowledge. As you listen, pay special attention to how the other groups are supporting their claims with science knowledge. Ask questions or make suggestions if you think a group's claim or recommendation is not as accurate or complete as it could be or if the group has not supported it well enough with evidence and science knowledge. Make sure you are convinced that each group's claims are accurate. Make sure you are convinced of their recommendations. If you are not convinced, share what you know about the body system they are making claims about, and try to offer a better claim, explanation, or recommendation.

Revise Your Explanations and Recommendations

As a class, work to make all of the claims and explanations as accurate and as convincing as possible. Then work to make all of the recommendations as complete and convincing as possible.

Update the *Project Board*

The final column of the *Project Board* asks you to summarize *What does it mean for the challenge or question?* The question for this Unit is *How can you prevent your good friends from getting sick?* It is time now to look at all the information you recorded on the *Project Board* and think about how what you now know helps you answer this question. What do you know, so far, that will help you answer the *Big Question?* Write your ideas in the last column. Add other questions you need to investigate. Write them in the *What do we need to investigate?* column.

nervous system: contains the brain, spinal cord, and an enormous number of nerves that carry messages about what is happening outside the body to the brain and messages from the brain to all parts of the body.

More to Learn

Other Body Systems

Your body has other systems than the ones you just read about. The systems you will read about now are just as important to your health and the way your body functions. All your body systems must work together to carry out the functions of life.

The Nervous System

Without your **nervous system**, you could not smell and taste your favorite foods or watch and listen to your favorite TV shows. Every breath you take and every beat of your heart depend on your nervous system. Your nervous system is an incredible communication system. It makes it possible for you to respond to things around you and to respond to what is taking place inside your body.

Your nervous system is made up of your brain, spinal cord, and an enormous number of nerves that form a network through your body. Nerves are very special cells. They carry messages about what is happening outside your body to your brain. They also carry messages about what is happening inside your body to your brain.

Your brain uses the information from these messages to coordinate how you will react. Then the brain sends messages using different nerve cells to tell your body parts what to do.

brain

spinal cord

peripheral nerves

There are a number of communicable diseases that affect the nervous system. You may have heard of **meningitis**. Meningitis is the inflammation of the lining surrounding the brain (called the **meninges**). Meningitis can be caused by viruses or bacteria. You also read about polio. Polio is a viral disease. The polio virus enters the nervous system and in some cases can cause paralysis. Vaccines have eliminated almost all cases of polio.

The Endocrine System

Your **endocrine system** controls the way other organs work. It is made up of **glands** (specialized tissue that secretes a substance) that produce and release **hormones**. Hormones are chemical messengers that regulate the activities of cells in an organ or group of organs. Hormones produced in organs of the endocrine system are carried through the body by the blood. This allows them to control even cells and organs that are far away from the glands that produce the hormones.

Your body could not operate without an endocrine system. Nearly everything that goes on inside your body is regulated in some way by the organs of your endocrine system. Your endocrine system regulates the amount of sugar in your blood, affects the rate at which food is converted into energy, and controls your growth rate. Without an endocrine system, your cells could not carry out the many functions they perform every day.

A major gland of the endocrine system is the **hypothalamus**, a tiny part of the brain near the middle of your head. The hypothalamus and the nervous system work together to control and regulate your body's responses to the environment, and to regulate growth, development, and reproduction. The hypothalamus is the link between the endocrine system and nervous system. It plays a major role in maintaining homeostasis (the process of keeping a constant internal environment) through the hormones and nerve messages it produces. You read earlier about homeostasis in the section on bacteria.

The endocrine system works by using a process called **feedback**. Feedback can be negative or positive. In negative feedback, when the amount of a particular hormone in the blood reaches a certain level, an endocrine organ sends signals that stop the release of that hormone. Positive feedback works in the opposite way.

meningitis: a viral or bacterial disease that results in inflammation of the lining surrounding the brain.

meninges: the linings (membranes) that cover and protect the brain and the spinal cord

endocrine system: the body system, made up of glands that produce hormones, which regulate all activities in organs and cells and help to maintain homeostasis in the body.

gland: specialized tissue that secretes a substance.

hormone: a chemical messenger that regulates the activities of cells in an organ or group of organs, or affects the activities of all the cells in the body.

hypothalamus: a major gland of the endocrine system that is part of the brain.

feedback: a process that uses nerve messages to turn on or turn off the production of hormones by the endocrine system.

PBIS

AIDS (acquired immune deficiency syndrome): a communicable disease caused by a virus that affects all the body systems.

wasting syndrome: a disease resulting from AIDS and affecting the hormones that metabolize food.

reproductive system: the body system that is specialized to produce sex cells (eggs and sperm) and sex hormones.

ovaries: female reproductive organs that produce eggs.

fallopian tube: a passageway for eggs from each ovary to the uterus.

uterus (or womb): hollow, muscular organ of the female reproductive system in which a fertilized egg develops.

fetus: a developing human from the ninth week of development until birth.

Most diseases of the endocrine system are non-communicable. However, **AIDS** (Acquired Immune Deficiency Syndrome), a communicable disease caused by a virus that affects the entire body, can also affect the endocrine system. AIDS can affect the production of hormones that regulate the metabolism of food. Without the proper hormones, food is not broken down properly, and the body does not receive the nutrition it needs. This condition can lead to **wasting syndrome**, a serious disease that is difficult to treat.

The Reproductive System

The **reproductive system** makes it possible for you to reproduce. Its organs are specialized to produce sex cells (eggs and sperm) and sex hormones. The organs in the reproductive system that produce sex hormones are also part of the endocrine system. The **ovaries** are the female reproductive organs that produce eggs. Each ovary is located near a **fallopian tube** that acts as a passageway for eggs as they travel to the **uterus** (womb). The uterus is a hollow, muscular organ where

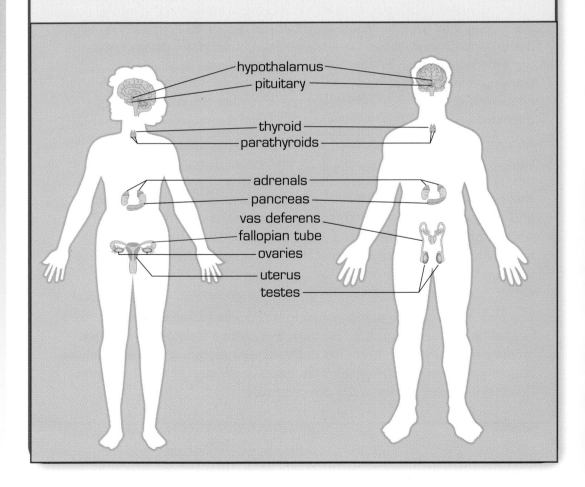

hypothalamus
pituitary
thyroid
parathyroids
adrenals
pancreas
vas deferens
fallopian tube
ovaries
uterus
testes

a **fetus** develops. The **testes** are the male reproductive organs that produce sperm. The sperm travel through tubes called **vas deferens** before they leave the body.

Many of the communicable diseases of the reproductive system are sexually transmitted diseases (STDs). One STD is AIDS. AIDS cannot be passed from person to person through casual contact. AIDS is sexually transmitted and also transmitted through blood transfusions or contact with other contaminated body fluids.

testes (singular, testis): male reproductive organs that produce sperm.

vas deferens: a tube that carries sperm from the testes in males.

The Excretory System

Your **excretory system** keeps the chemicals in your body in balance. The food you eat, the water you drink, and the air you breathe provide all the materials your body needs to survive. The materials are broken down inside your body by the digestive system into the proper chemicals that are sent through your blood to your cells. Your cells use these chemicals, converting them into energy and other substances used for growth and survival.

The process of changing food, water, and air into useful chemicals produces waste materials that can't be used by your body. Nitrogen compounds, salts, carbon dioxide, and excess water are all waste products. If these materials remain inside your cells or blood, the build-up of harmful wastes could result in disease and damage your body. It is the job of your excretory system to get rid of the harmful waste materials.

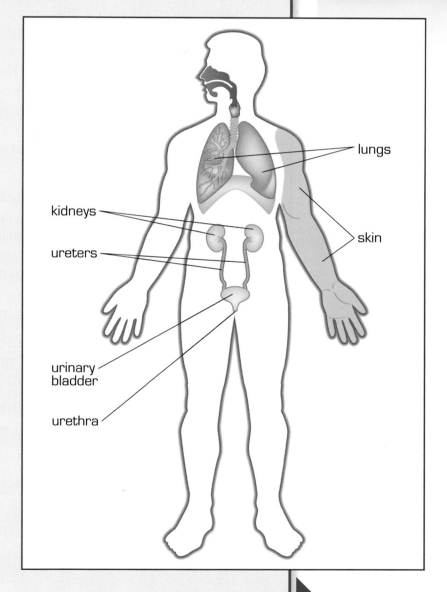

lungs

skin

kidneys

ureters

urinary bladder

urethra

kidney: one of two excretory organs that filter the blood and remove waste products.

urine: a watery fluid produced by the kidneys that contain wastes.

ureter: one of two tubes that carries urine from each of the kidneys to the urinary bladder.

urinary bladder: an organ that stores urine until it is eliminated from the body.

urethra: a tube through which urine flows from the body.

sweat glands: structures in the skin that remove water and salts.

cystitis: an infection of the lining of the urinary bladder, caused by bacteria.

skeletal system: the internal framework of bones that provides shape and support, allows movement, protects organs, produces blood cells, and stores minerals.

marrow: soft tissue inside the bones that makes red blood cells and stores fat.

Your excretory system is made up of several organs. **Kidneys** filter your blood and remove the waste products of metabolism. These waste products are combined with water to form **urine**, which moves through your **ureters** to your **urinary bladder** and out of your body through your **urethra**. Most of the waste is removed through your kidneys. As your kidneys filter the blood, they help maintain homeostasis by regulating the amount of water in your body.

Your lungs and skin also remove some wastes from your body. When you exhale, carbon dioxide and some water are removed from your body. **Sweat glands** in your skin also remove water and some salts. Some wastes need to be broken down before they can be removed. Your liver carries out this function.

One common infectious disease of the excretory system, **cystitis**, is caused by a bacterial infection. Cystitis affects the lining of the urinary bladder and makes elimination of urine painful. Cystitis is easily treated with antibiotics, which you read about earlier in **Learning Set 2**.

The Skeletal System

When you look up at a tall building, you may wonder how it keeps from falling down. The building has an internal structure of steel or wooden beams that act as a framework. The framework is not visible, but it is necessary to keep the building from collapsing. Your body also has an internal framework that you cannot see. This framework is your **skeletal system** and is made up of the bones in your body. Without your bones, you could not stand up, and your body would collapse.

Your skeletal system has many functions other than support. It allows you to move, protects your organs, produces blood cells, and stores minerals and other materials. A special material, called **marrow**, is found inside your bones. One type of marrow makes red blood cells that carry oxygen to the cells in the body. Without bones, not only would your body collapse, but your cells would be unable to get the oxygen they need to survive. The other type of marrow stores fat that can act as an energy reserve.

You probably already know about your bones acting as support and protection for your body, but you probably never thought about the fact that your bones are living tissue. Bones are not only strong, they contain living cells. They grow and develop, just like the other parts of your body. As you grow, new bone tissue grows. Blood vessels and nerves enter your bones through small canals to the living cells inside and allow the bones to grow. Because the cells of your bones receive nerve impulses and blood that contains nutrients and oxygen, bone can also heal itself when it is broken. New bone tissue grows to fill the gap between the broken ends of the bone.

Osteomyelitis is a bone disease caused by bacteria or fungi that is often in another part of the body and spreads to the bone through the blood. The bone disease is not communicable, but the bacteria that cause it can be passed from person to person

The Muscular System

You may ride a bike, play sports, or walk to school. These are all active movements. But you also breathe, swallow, smile, and sit in a chair. You digest food, and your heart beats. Without a **muscular system,** you would be unable to do any of these things.

Your muscular system is made up of the muscles in your body. Like other body tissues, muscle tissue is living tissue that receives blood and nerve impulses. The muscular system works together with the nervous system, the skeletal system, the circulatory system, and your other body systems to move the parts of your body. Some of the things you do, like riding a bike and walking, are voluntary movements. They are under your conscious control and are controlled by **voluntary muscles**. Other things, like breathing and the beating of your heart, are involuntary. **Involuntary muscles** are not under your conscious control.

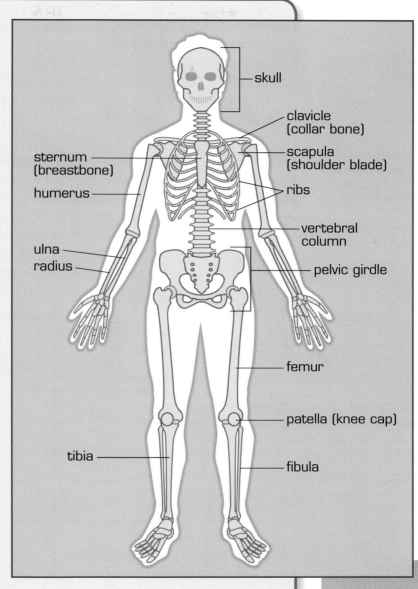

skull
clavicle (collar bone)
scapula (shoulder blade)
sternum (breastbone)
ribs
humerus
vertebral column
ulna
radius
pelvic girdle
femur
patella (knee cap)
tibia
fibula

osteomyelitis: an infectious bone disease caused by bacteria or fungi.

muscular system: the body system, made up of the muscles, which moves the parts of the body.

voluntary muscles: muscles that can be moved by conscious control.

involuntary muscles: muscles that move without conscious control.

Muscles move by contracting; they become shorter and thicker. The muscle cells contract when they receive messages from your nervous system. Because muscle cells can only contract, not extend, they have to work in pairs. When one muscle contracts, the other muscle in the pair relaxes to its original length. The bones of the skeletal system support the muscles. With your skeletal system and your muscular system working together, you are able to move your body.

Polio, a viral disease you read about earlier, affects the muscular system as well as the nervous system. When the polio virus enters the nervous system, the nerve cells can become inflamed. This inflammation leads to destruction of the nerves that supply the muscles. As a result, the muscles become weak and poorly controlled. Finally, the muscles can become paralyzed.

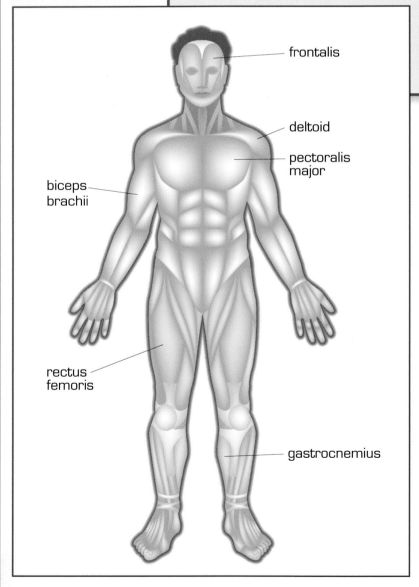

frontalis

deltoid

pectoralis major

biceps brachii

rectus femoris

gastrocnemius

Learning Set 4

How Do Scientists Identify and Stop Disease Outbreaks?

You read about many different diseases, the bacteria and viruses that cause them, and their symptoms. You read about how body systems work and how they work together to keep you healthy, to fight disease, and even to spread disease. You simulated how bacteria and viruses can be transmitted to others and read about how diseases can spread. You have seen many examples showing how hard it is sometimes to keep from spreading diseases.

Bacteria and viruses are responsible for most communicable diseases. It is amazing that there are so many different diseases and ways to spread them. One reason for this is that bacteria and viruses change to react to changes in the environment.

When this happens, scientists have to learn how to combat the new microbes and the diseases they cause. Several recent **outbreaks** of new viral diseases have surprised scientists. Each time, however, these new viruses are stopped from spreading further by actions taken by people to prevent their spread.

In this *Learning Set*, you will investigate how scientists identify and stop disease outbreaks.

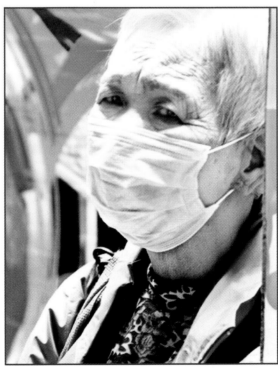

This person is protecting herself from the SARS virus (Severe Acute Respiratory Syndrome) in 2003. What is she doing to prevent getting sick or spreading the disease to someone else?

outbreak: occurrence of more cases of a particular disease than expected in a given period of time or a given area.

4.1 Understand the Question

Thinking About Disease Outbreaks

The question for this *Learning Set* is *How do scientists identify and stop disease outbreaks?* It is a good idea to think about what you know about disease outbreaks. It is also important to think about what you are unsure about and what you would like to investigate.

Sometimes, bacterial or viral diseases start, and no one predicted they would happen. When people got *E. coli* from eating contaminated spinach, nobody could have predicted it. That's because nobody knew the spinach was contaminated. The same thing happened in 2003 when the SARS (Severe Acute Respiratory Syndrome) virus made many people around the world very sick. Nobody knew where it came from. When many people get sick unexpectedly, it is called an outbreak. Outbreaks can happen for many reasons. SARS was caused by sudden change, or **mutation**, in a virus. When a virus mutates, its shape changes. Then the antibodies that used to protect against it no longer work.

mutation: a permanent change in the genetic material of a cell.

Get Started

In your group, share your ideas about what you think you know about disease outbreaks. Listen carefully to all the ideas presented. Consider what you know about

- how scientists find out about disease outbreaks, and

- why it is important to investigate outbreaks.

Think of several questions that might help you answer the question for this *Learning Set*. Develop two questions that might help you better understand disease outbreaks. Then share your two questions with your small group. Carefully consider each question and decide if it meets the criteria for a good question. With your group, refine the questions that do not meet the criteria. Choose the two most interesting questions to share with the class. Give your teacher the rest of the questions so they might be used later.

Update the *Project Board*

Share with the class your group's ideas about disease outbreaks. You can record these ideas in the *What do we think we know?* column. Next, share with the class your group's two questions. Your teacher will add your questions to the *Project Board*. Throughout this *Learning Set*, you will work to answer some of these questions.

4.2 Read

How Can Disease Outbreaks Be Identified?

You have read about the contaminated spinach that killed a number of people in the United States in 2006. The spinach had been sprayed with water containing *E. coli*. The water used to irrigate the spinach fields had accidentally been contaminated by animal feces from a nearby farm.

It is important to stop outbreaks of serious diseases before they cause a lot of damage. Sometimes a disease can be **contained** by identifying the source (how the disease began) and eliminating it. Finding the source of a disease can be complicated. Epidemiologists had to make many discoveries to find the source of the *E. coli* outbreak you read about. The first discovery revealed that everybody who had the disease had eaten spinach. Scientists then discovered what fields the contaminated spinach came from. Finally, it became clear how the spinach had become contaminated, through animal feces in the water used for irrigation.

All communicable diseases cause discomfort. Even mild diseases can cause serious problems, even death, especially in babies, elderly people, and people already sick with something else. To contain a communicable disease, it is important to keep infected people from infecting others. People who are contagious are kept in **quarantine**; they must stay in one place and are not allowed near healthy people. It is not always possible to quarantine every contagious person, especially when a disease has a long **incubation period**, the amount of time between exposure to a disease and when symptoms begin. During this time, a person is still contagious and can spread the disease to others, even though they show no symptoms. When incubation periods are long, many people will already be infected by the time an outbreak is discovered.

Outbreaks of disease in people can also be caused by disease in animals. Avian influenza is commonly called bird flu. Outbreaks occur worldwide in poultry from time to time. Some types of this flu virus can cause a severe and very contagious illness in birds and can be fatal to the birds. Some types of the virus can also infect people. People do not get bird flu from eating chicken. To catch bird flu, a person must be in close, daily contact with infected birds. Farmers who raise chickens are usually the people who become infected with bird flu.

contained: keeping a disease from spreading.

quarantine: keeping infected people away from healthy people.

incubation period: the amount of time someone may be infected with the disease before he or she shows symptoms. During this time, an infected person is contagious.

Chickens can be carriers of avian influenza, or bird flu.

Read the news release from the Public Health Service for Wales.
As you read the news release, think about how the disease spread.
Then think about how the disease can be tracked and contained.

Human cases of avian influenza A/H7N2 in the United Kingdom

Following the confirmation on 25 May 2007 by Health Authorities of the United Kingdom, of influenza A/H7N2 virus infection in four individuals (two in Wales and two in northwest England) exposed to infected poultry at small-holding [small farm] Corwen Farm, Conwy, Wales, the National Public Health Service (NPHS) for Wales is continuing with the investigation of the incident and with the implementation [putting into action] of public health measures.

As of 30 May 2007, 256 individuals, exposed either to affected premises [buildings], infected poultry, or to another individual with confirmed or presumptive [assumed to have the flu] influenza A/H7N2 virus infection, were identified in the following settings: household (39), school (14), and workplace

(203), including, at least, 148 patients and staff at two hospitals. Seventy-nine (79) of the exposed individuals are no longer considered at risk as the 7-day incubation period has elapsed. As a precautionary measure, and in accordance with UK policy, the decision was made to offer oseltamivir [a medicine to treat this type of flu] to exposed individuals.

In addition to the two Wales residents with laboratory confirmed infection, 17 of the exposed individuals, including one healthcare worker, present [show] or have presented with influenza-like illness (fever above 38°C, aches and pains, cough/head cold, sore throat or conjunctivitis [eye infection]). None of them is seriously ill and they are receiving and have received oseltamivir treatment.

Although, according to the preliminary results of the epidemiological investigation [scientific study of the disease], limited human-to-human transmission cannot be ruled out, the public health risk is considered low.

From www.euro.who.int/flu/situation/20070601_1
(World Health Organization Regional Office for Europe)

Stop and Think

1. What are the different ways avian influenza (bird flu) can be spread?

2. Suppose that a person exposed to the bird flu does not develop any symptoms for six days after exposure. Describe why you cannot say for sure that the individual does not have the bird flu.

3. What are ways you can think of to keep from spreading bird flu?

What's the Point?

To track disease outbreaks, scientists need to identify the source of the disease or how it began. They must also keep track of who was exposed to the disease and who developed symptoms. This is particularly difficult if the disease has a long incubation period. To keep the disease contained, it is important to keep infected persons from infecting others.

4.3 Investigate

Track a Disease

When an outbreak happens, epidemiologists do two things: They make recommendations about how to keep the disease from spreading, and they try to trace where the disease began. If they can do both of these things, there is a good chance they can stop the outbreak and avoid further infection. It is important to understand the causes of an outbreak in order to stop it.

Like you did in your simulation of disease spread, epidemiologists trace the spread of an infection *backward* to determine where the infection began. They use a lot of different information to do that.

In the next two short investigations, you are going to trace the path of a disease just like an epidemiologist might. You will not be able to do everything an epidemiologist does, but these investigations will help you understand the kind of work epidemiologists do and might give you some ideas about how to prevent the spread of disease in your school.

Procedure: Tracking Disease Through Simple Interactions

Follow the interactions of the people who are described below to determine who made the other people sick.

Joe had breakfast with Peter and Mike. Joe then went to class with Sally. Sally and Peter went to the movies. The next day, Joe, Sally and Peter were sick.

1. To trace the interactions, sketch a diagram that shows how the disease spread through the group. Use each person's name and connect it with an arrow to the person with whom they interacted. One set of interactions has been done to get you started. Complete the interactions connecting Joe to Sally and Sally to Peter. Circle the names of the people who got sick.

2. The first person to get sick in a population is called the initial or sentinel case. Use the diagram you sketched to decide who was the sentinel case, the person who got the others sick.

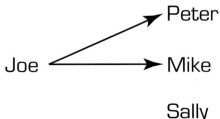

3. What information did you use to figure this out? Did you make any mistakes along the way? If you did, what caused you to change your mind?

Communicate Your Results

Together with the class, sketch the interaction diagram and discuss how you figured out who was the sentinel case. Discuss what information you used to figure it out. Share with the class any mistakes you made along the way and what made you change your mind.

Procedure: Tracking Disease Through More Complex Interactions

The interactions you just investigated were pretty simple. Usually, the interactions among people are more complicated. Remember the investigation you did at the beginning of the Unit when you shared liquid with others as a way of interacting with them? The interactions were hard to remember after they happened. You could remember them well only if you wrote them down as they were happening. In the real world, you interact with a lot of people. You interact directly with people. You shake hands. Others might sneeze or cough on you. You also interact indirectly with other people when you touch things that others have touched. When you are sick, you can spread germs either directly or indirectly.

The interactions among this next population are more complicated than the interactions in the first example. In this case, Kristin starts the interactions. The last interaction is from David to Kristin. The diagram of interactions is shown on the next page. Your job is to describe the interactions and to imagine how they might have happened. Kristin, Sue, and David all got the same illness within a few hours of each other. Kelly and Bill did not get sick.

1. Write a set of sentences that describes all the interactions shown in the diagram. Use sentences from the exercise above to help you. Write your sentences as a simple story about everyday activities. Imagine what the people were doing and why they were interacting with each other.

2. Now, try to figure out who the sentinel case is in this situation. Notice who got sick and who did not get sick. Make sure your answer can explain both.

   - Who was the sentinel case?

   - How do you know this?

   - Why do you suppose that Kelly or Bill did not get sick?

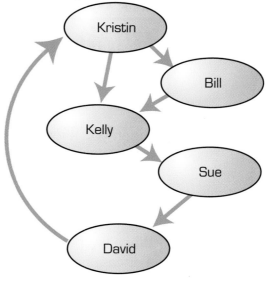

Communicate Your Results

This problem was more difficult to solve, and your group may have disagreed about the correct answer. Share the different answers, compare all the answers to the evidence provided, and figure out who was the sentinel case.

- Why is this the best answer? What makes it better than other answers your classmates came up with?

- What made this investigation harder than the first one?

What's the Point?

When an outbreak of a disease occurs, epidemiologists need to trace where the disease began and make recommendations to stop the spread of disease. Tracing where a disease began can be very tricky, especially when a disease has a long incubation period. Tracking where a disease came from requires tracing the spread of the disease backwards from those who are sick to identify what caused the disease. Epidemiologists must explain why some people got sick and why others did not.

4.4 Read

What Makes Disease Tracking Hard?

When you know everything about people's interactions, you can figure out where a disease started and how to keep it from spreading. You were able to figure out the sentinel case in the previous investigation by thinking about alternatives and which one explained the situation best. But you had to think about only a few alternatives.

When the population is larger and you cannot track every interaction, or when a disease has a long incubation period, it gets a lot harder. There are many more possibilities that need to be considered.

When you also add travel into the picture, it gets even harder to track diseases and stop them. The influenza virus of 1918 spread around the world. This was during World War I, and the virus was taken to distant places by soldiers. Scientists know how it was spread around the world, but they still do not know how it got started. When the SARS outbreak happened in 2003, people flying to and from China for business and vacations spread the virus around the world.

Soldiers may have unknowingly spread diseases to different parts of the world.

When an **epidemic** spreads throughout the world, it is called a **pandemic**. Pandemics have ancient roots. Several documented pandemics affected the outcome of wars. In the 16th century (1500s), the Aztec population was beaten in war by the Conquistadors of Spain. The Aztecs lost even though they had more warriors and a bigger population. Unfortunately, the Spaniards brought smallpox with them to Mexico. The Aztecs were not immune to that disease, and many Aztecs got sick or died.

> **epidemic:** a rapidly spreading outbreak of an infectious disease.
>
> **pandemic:** an epidemic that has spread throughout the world.

Diseases spread very quickly in our world today. When people lived in rural places, like farms and small towns, they did not have contact with many people. Also, people didn't travel much outside their towns. Because there were few interactions outside of one's small town, diseases did not have a chance to spread outside of communities.

As cities have grown and more people have moved to them, diseases spread more easily than they used to. There are several reasons for this. In many large cities, in the USA and around the world, sanitation is still a problem. When there is dirt, waste, or sewage in the streets, a variety of diseases can live and spread to people. Also, when there is poor sanitation water sometimes becomes contaminated with diseases that spread to people who use the water. All these things make people sick. Then because of the number of people in a city, diseases can spread quickly because there are so many interactions among people.

Diseases spread more easily in crowded areas.

GOOD FRIENDS and GERMS

Finding ways to prevent the spread of disease is important. If diseases can be stopped before many people get sick, or if diseases cannot be spread from person to person, fewer people will get sick.

Scientists continually work to find ways to minimize outbreaks and prevent the spread of disease. Governments around the world have methods for reporting the incidence of disease outbreaks. Your doctor and school are required to report outbreaks of some communicable diseases to the health department where you live. When incidents of a disease are reported to a central place, scientists can identify outbreaks at an early stage and immediately get to work warning people about what they need to do so they won't catch it. For example, bacterial meningitis is highly contagious and very dangerous. When even one case occurs at a school or college, the school contacts everybody who works or goes to school in that place and tells everybody who has come in close contact with the sick person to be checked for infection. Those who are infected immediately get antibiotics. Even those who are not infected often get antibiotics to prevent them from getting the disease. The earlier this disease is treated with antibiotics, the easier it is to cure.

Stop and Think

1. Why does a long incubation period make it difficult to track a disease?

2. Why do diseases spread more quickly today than they did many years ago?

3. Doctors are required to report cases of some highly contagious diseases. Why do you think they need to do that?

What's the Point?

You have seen that tracking the interactions among small groups of people is sometimes difficult. Imagine how hard it would be to track the interactions of people when groups get very large. It gets even harder to track interactions when groups interact with each other. When it is hard to track interactions, it is also hard to track the spread of disease and to figure out how it got started. When a disease has an incubation period of more than a day or two, it becomes even harder to figure out how a disease got started, because it is practically impossible, when a population is large, to track who interacted with whom over a long period of time. Nonetheless, to stop a disease from spreading further, it is important to find out its source. Health officials need to find out who was the first, or sentinel, case and how that person got the disease.

Learning Set 4

Back to the Big Question

How can you prevent your good friends from getting sick?

The *Big Question* for this Unit is *How can you prevent your good friends from getting sick?* To help you answer this question, you investigated how scientists identify and stop disease outbreaks. You read about how tracking interactions among large groups of people is very difficult. You also experienced that it can be extremely difficult to find out who was the first person to get a disease.

Explain

Earlier, your explanations focused on how diseases are passed between individual people. In this *Learning Set*, you have focused on how diseases are passed around populations and between populations. Before going back to the *Big Question*, you will use that knowledge to explain two more things.

- How do diseases spread through populations?

- How do diseases spread across populations?

As you've done previously, use a separate *Create Your Explanation* page for your answer to each question. Use your answer as your claim. Then fill in evidence and science knowledge from this *Learning Set* that supports your claim. Then develop your explanation—a logical statement that ties together your claim and your evidence and science knowledge that makes your claim convincing. Be prepared to share your claims and explanations with the class.

Recommend

As you have experienced throughout this Unit, your claims and explanations about how people get sick are useful in developing recommendations about staying well. Use the claims and explanations you have just developed to come up with recommendations about helping populations avoid contagious diseases. Use a separate *Create Your Explanation* page for each of your recommendations, record evidence and science knowledge that supports each, and develop an explanation for each. Make sure your recommendations answer the following question:

- How can the public be protected from contagious diseases?

Also, look back at your old recommendations, and see if any of those should be revised. Revise those that you now know more about. Be prepared to share your new recommendations with the class.

Update the *Project Board*

The *What are we learning?* column on the *Project Board* helps you pull together everything you learned. Remember always to include your evidence. You can then use what you have learned to answer the *Big Question*.

Each investigation you do is like a piece of a puzzle. You must fit together the pieces to help you address the challenge. Your *Big Question* was *How can you prevent your good friends from getting sick?* The last column, *What does it mean for the challenge or question?* is the place to record how learning about bacteria and viruses can help you answer the *Big Question*.

How can you prevent good friends from getting sick?				
What do we think we know?	**What do we need to investigate?**	**What are we learning?**	**What is our evidence?**	**What does it mean for the challenge or question?**

Answer the Big Question

How Can You Prevent Your Good Friends from Getting Sick?

It is now time to pull all your learning together. First, each group will focus on one communicable disease. They will describe the disease and how to prevent that disease from being spread to other people. Then they will make recommendations about what can be done to avoid transmitting the disease to others. When finished, each group will present its findings to the class in a *Solution Showcase*. Then you will make posters with recommendations for helping to keep the school healthy.

Describe Your Disease

Your first task is to describe enough about the disease you are assigned to make recommendations about how to prevent its spread. Use the questions on the next page to guide you in putting together a presentation about your disease. It is important to answer all of the questions. Some of them you already know the answers to. Some you might have to figure out based on what you have read and done. The explanations and recommendations you have been working on throughout the Unit will also help you answer the questions. You might also need to do some additional reading.

As you decide how much information to provide in answering each question, remember that your goal is to make recommendations about how to avoid spreading the disease to others. You need to answer each of the questions well enough so your classmates will trust statements you make about how to prevent the disease from being spread.

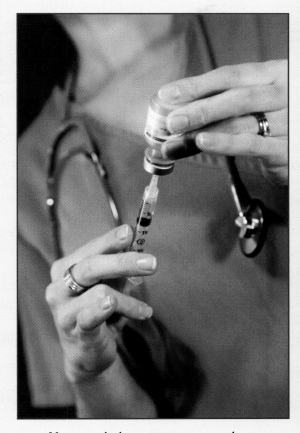

Vaccines help protect against diseases.

Symptoms of the Disease

- What are the symptoms of the disease?

- Which symptoms would help a healthcare worker diagnose the disease in a person? Which symptoms are unique to this disease?

- How sick do you get with this disease? How long are you sick? Can the disease kill a person?

Cause

- Is the infection bacterial or viral? Describe in detail what caused the disease.

- How does the bacteria or virus interact with the body to produce the disease?

- What cells or tissues does it infect, and how does it damage those cells or tissues?

Transmission

- How is this disease transmitted from person to person?

- What are the best ways to prevent this disease from spreading? How can people prevent this disease from being transmitted from one person to another?

Epidemiology

- Where did this disease first emerge?

- Is it spceific to a particual place or spread over a lager area? Describe the spread of this disease through a population or around the world.

- What human actions spread this disease?

- What human actions can prevent its spread?

Treatment

- How is this disease treated?

- Is there a vaccine available? If not, why is there no vaccine?

- Can antibiotics work against this disease? Explain why antibiotics can, or cannot, be used against this disease.

- How will this disease affect humans in the future? Will it be eradicated, or will it remain a disease that affects humans?

Plan Your Solution

Make Your Recommendations

Now, recommend ways of preventing the spread of your disease. Remember from other PBIS Units that the best recommendations are supported by evidence and science knowledge.

Try to make three types of recommendations:

- What do you recommend people do to avoid getting the disease?

- What do you recommend people do to avoid spreading the disease?

- What do you recommend populations do to avoid spreading the disease?

Use a *Create Your Explanation* page for each recommendation. Record your recommendation as a claim. Then provide evidence and science knowledge to support your claim. For each, try to include reasons why others should follow your recommendation. Your explanation should tie your claim, evidence, and science knowledge together and include the ways the disease is spread.

Communicate Your Solution

Solution Showcase

Make a poster that reports on your disease. Tell your class about the disease you are reporting about. Be sure to be especially clear about how the disease interacts with different body systems, how it is transmitted, the human actions that cause it to spread, and ways to prevent its spread. Your classmates will be particularly interested in understanding enough about your disease to know why the ways of preventing its spread will work. Make sure you answer the following questions in your presentation:

- What is the disease?

- What are its symptoms?

- What body systems does it affect?

- How does it spread to others?

- What particular things do people do that help spread this disease?

- How is it treated?

- How can its spread be prevented?

- What do you recommend people do to avoid getting the disease?

- What do you recommend people do to avoid transmitting it to others?

- What do you recommend populations do to combat this disease?

- For each one of your recommendations, explain why it would work.

As you listen to your classmates make their presentations, make sure you understand how to prevent the disease from spreading and how well each of their recommendations will work. If you do not understand those things, or if you do not think their recommendations will work, ask questions. The questions above might help you figure out what questions to ask.

Update the *Project Board*

The important thing to focus on in updating the *Project Board* this time is applying what you know to the *Big Question*. This goes in the last column. The *Big Question* is about how you can prevent your good friends from getting sick. What have you just learned from these presentations that can help you answer that question? In the last part of this Unit, you will be writing recommendations for making your classroom or school a healthier place. Update the *Project Board* with this in mind.

Make Your School a Healthier Place

Now that you know some things about how to prevent the spread of many different bacterial and viral illnesses, you can make recommendations about how to make your classroom or school a healthier place. You will certainly want to recommend things like sneezing into a tissue and washing your hands after you sneeze. But this Unit has a lot more than that in it about how to prevent the spread of diseases.

Your class's final product will be a set of recommendations for staying healthy and helping others stay healthy. Each recommendation will be supported by evidence from your investigations and science knowledge you learned.

Each group in the class will play a part in contributing to that list. It will be important to do several things in preparation for your group work:

- As a class, identify which diseases you will make recommendations about.

- As a class, create one recommendation together. Begin by identifying explanations and recommendations you've already made that are relevant to making a recommendation about this disease. From those explanations and recommendations, develop a more specific recommendation about how to prevent the spread of that disease in your school. Attach evidence from your investigations and science knowledge from your readings to support it, and develop an explanation statement to go with it. The explanations and recommendations you have developed during this Unit should help you with this.

- As a class, review the last column on the *Project Board*, the one that suggests answers to the *Big Question*.

Then each group will make recommendations about preventing the spread of disease in some place in the school—the classroom, the lunchroom, the bathroom, and so on. Use a Create Your Explanation page for each of your recommendations.

Make a poster that can be displayed in the school. Your poster should convince others in your school to follow your advice.

English & Spanish Glossary

A

AIDS Acquired Immune Deficiency Syndrome, a communicable disease caused by a virus that affects all the body systems.`

SIDA Síndrome de Inmunodeficiencia Adquirida, una enfermedad contagiosa causada por un virus que afecta todos los sistemas del cuerpo humano.

antibiotic A type of medicine that kills microbes or prevents them from growing.

antibiótico Un tipo de medicina que mata microbios microorganismos o evita que éstos crezcan.

antibodies Substances that help destroy harmful microbes.

anticuerpos Sustancias que ayudan a destruir microbios nocivos.

anus The bottom opening of the digestive tract.`

ano La abertura al final del tracto digestivo.

artery A blood vessel that carries blood away from the heart.

arteria Un vaso sanguíneo que lleva sangre fuera del corazón.

atria (singular, atrium) Heart chambers that receive blood from the veins.

aurículas (en singular aurícula) Cámaras del corazón que reciben sangre de las venas

B

bacteria (singular, bacterium) The most common form of one-celled organisms.

bacterias (singular, bacteriaum) La forma más común de los organismos unicelulares.

barrier Something that obstructs (or blocks).

barrera Algo que obstruye (o bloquea).

blood vessels Tube-like parts of the circulatory system that transport blood through the body.

vasos sanguíneos Partes del sistema circulatorio en forma de tubos que transportan sangre a través del cuerpo.

bone marrow Soft tissue in the center of bones where most blood cells are produced.

médula ósea Un tejido suave en el centro de los huesos donde se producen la mayoría de las células sanguíneas.

bull's-eye A circular spot, usually black, at the center of a target. The bull's-eye is surrounded by concentric circles.

blanco Un punto circular, generalmente negro, en el centro de un objetivo. El blanco está rodeado de círculos concéntricos.

C

calorie The amount of energy needed to raise the temperature of one gram of water by 1°C.
caloría La cantidad de energía necesaria para aumentar la temperatura de un gramo de agua por 10C.

Calorie The amount of energy in foods. One Calorie is the same as 1 kilocalorie or 1000 calories.
Caloría La cantidad de energía en los alimentos. Una caloría es igual a una kilocaloría ó 1000 calorías.

carbohydrate A type of substance found in food that gives the body the energy it needs.
carbohidrato un tipo de sustancia que se encuentra en los alimentos que le provee al cuerpo la energía que éste necesita.

carrier A person who has a contagious disease.
portador Una persona quien tiene una enfermedad contagiosa.

case study An observation of a person or group to use as a model.
estudio de caso Una observación de una persona o grupo para ser usado como modelo.

cell The structural and functional unit of all living organisms. (What a living thing is made of and what makes it work.) It is sometimes called the "building block" of life.
célula La unidad estructural y funcional de todos los organismos vivos. (De qué está compuesto un organismo vivo y lo que lo hace funcionar). Algunas veces se le conoce como el "componente básico" de la vida.

cell membrane A structure that surrounds the cell; controls the movement of materials into and out of the cell.
membrana celular Circunda la célula y controla el movimiento de materiales que entran y salen de de la célula.

cell theory A theory about the relationship between cells and living things.
teoría de la célula Una teoría acerca de la relación entre las células y los objetos vivientes.

cellular respiration The process by which food and oxygen are converted to energy, carbon dioxide, and water.
respiración celular El proceso por el cual el alimento y el oxígeno son convertidos en energía, bióxido de carbono y agua.

English & Spanish Glossary

cell wall Protects and supports the plant cell.

pared celular Protege y apoya la célula vegetal.

chlorophyll A green pigment that makes photosynthesis possible in plants.

clorofila Un pigmento verde.

chloroplast An organelle that contains the green pigment chlorophyll used in photosynthesis.

cloroplasto Organelo que contiene el pigmento verde conocido como clorofila usado en la fotosíntesis.

chromosomes The structures containing the genetic material of the cell.

cromosomas Contienen el material genético de la célula.

circulatory system An organ system that carries nutrients and other chemicals to the cells of the body and carries away waste; includes the heart, arteries and veins.

sistema circulatorio Un sistema de órganos que lleva nutrientes y otros químicos a las células del cuerpo y saca desperdicios; incluye el corazón, las arterias y las venas.

colony a group of only one kind of bacteria that grow from a single, original bacterium.

colonia Un grupo de solamente un tipo de bacteria que crece de una bacteria original simple.

communicable disease A disease that spreads very readily from person to person.

enfermedad contagiosa Una enfermedad que se transmite fácilmente de una persona a otra.

compound light microscope A microscope that has more than one lens and uses light transmitted to your eye to form an image.

microscopio de luz compuesto Un microscopio que tiene más de un lente y usa la luz transmitida a tu ojo para formar una imagen.

contained Keeping a disease from spreading.

detener Evitar la propagación.

contract To shorten.

contraer Acortar.

cystitis An infection of the lining of the urinary bladder, caused by bacteria.

cistitis Una infección en el recubrimiento de la vejiga urinaria, causado por bacteria.

cytoplasm The watery fluid that contains the organelles of the cell.
citoplasma El fluido acuoso que contiene los organelos de la célula.

D

dehydration A condition in which the body does not have enough fluid to function properly.
deshidratación (médica) Una condición en la cual el cuerpo no tiene suficientes fluidos para funcionar adecuadamente.

diameter The length of the line through the middle of a circle from one side to the other.
diámetro La longitud de la línea a través del centro de un círculo de un lado a otro.

diaphragm A band of muscle that regulates the pressure in the chest cavity.
diafragma Una banda de músculos que regula la presión en la cavidad del pecho.

diarrhea Frequent and watery bowel movements.
diarrea Movimientos frecuentes y acuosos de los intestinos.

disinfectant Substance that can kill microorganisms.
desinfectante Sustancia que puede matar microorganismos.

dormant Temporarily not active.
durmiente Temporalmente inactivo.

droplet transmission A way that an infectious disease can be transmitted. Droplets containing an infectious agent (bacteria or viruses) are released into the air when an infected person sneezes, coughs, talks, or exhales. They then come into contact with another person's eye, nose, or mouth.
transmisión de gotitas Una maneta en la cual una enfermedad contagiosa puede ser transmitida. Las gotitas que contiene un agente infectado (bacteria o virus) son liberadas en el aire cuando una persona infectada estornuda, tose, habla o exhala. Éstas entran en contacto con los ojos, nariz o la boca de otra persona.

E

endocrine system The body system, made up of glands that produce hormones, which regulate all activities in organs and cells and help to maintain homeostasis in the body.
sistema endocrino El sistema corporal hecho de glándulas que producen hormonas, el cual regula todas las actividades en los órganos y células y ayuda a mantener la homeostasis en el cuerpo.

English & Spanish Glossary

enteric bacteria Bacteria that live in the digestive tract.

bacteria entérica Bacteria que vive en el tracto gastrointestinal.

enzyme A substance that causes a chemical change in another substance.

enzima Una sustancia que causa un cambio químico en otra sustancia.

epidemic A rapidly spreading outbreak of an infectious disease.

epidemia Una brote con rápido esparcimiento de una enfermedad contagiosa.

epidemiologist A scientist who studies diseases and how they spread.

epidemiólogo Un científico que estudia enfermedades y su cómo se propagan.

epidemiology The study of how and why diseases occur.

epidemiología El estudio de cómo y por qué ocurren las enfermedades.

eradicated Wiped out.

erradicado Eliminado.

esophagus The tube that carries food from the mouth to the stomach.

esófago El tubo que lleva alimento de la boca al estómago.

eukaryote An organism whose cells have nuclei.

eucariota Un organismo cuyas células tienen núcleos.

expel to force out.

expulsado Obligado a salir.

F

fallopian tube A passageway for eggs from each ovary to the uterus.

trompa de falopio Un conducto para los óvulos de cada ovario hacia el útero.

fat A type of substance found in food that gives the body the energy it needs.

grasa Un tipo de sustancia que se encuentra en los alimentos que le provee al cuerpo la energía que éste necesita.

fetus A developing human from the ninth week of development until birth.

feto Un ser humano en desarrollo desde la novena semana hasta su nacimiento.

feces Waste that is produced in the digestive system.

excrementos Desperdicio que es producido en el sistema digestivo.

feedback A process that uses nerve messages to turn on or turn off the production of hormones by the endocrine system.

reacción Un proceso que usa mensajes nerviosos para activar o desactivar la producción de hormonas por el sistema endocrino.

field of view The circle of light you see when you look through a microscope.

campo de visión El círculo de luz que ves cuando miras a través de un microscopio.

filter To separate.

filtrar Separar.

flagellum (plural, flagella) A string-like part on some cells that helps the cell move.

flagelo (plural, flagelos) Una parte en forma de cuerda en algunas células que ayuda a que la célula se mueva.

fungi (singular, fungus) Eukaryotes that have cell walls, reproduce by spores, and get food by absorbing it from their surroundings.

hongos Las eucariotas que tienen paredes celulares, se reproduce por esporas y obtiene alimento absorbiéndolo de sus alrededores.

G

gastrointestinal Related to the stomach and intestines.

gastrointestinal Relacionado con el estómago y los intestinos.

genetic material Contains the information that determines the traits of an organism; hereditary material.

material genético Contiene la información que determina los rasgos de un organismo; material hereditario.

gland Specialized tissue that secretes a substance.

glándula Tejido especializado que segrega una sustancia.

H

heart The organ responsible for pumping blood through the blood vessels of the body.

corazón El órgano responsable de bombear sangre a través de los vasos sanguíneos del cuerpo.

heart chamber One of four parts — right atrium, right ventricle, left atrium, and left ventricle — of a heart.

cámara del corazón Una de cuatro partes –atrio derecho, ventrículo derecho, atrio izquierdo, y ventrículo izquierdo– de un corazón.

English & Spanish Glossary

homeostasis The process by which an organism keeps its internal environment in a constant condition despite changes in its external environment.
homeostasis El proceso por el cual un organismo mantiene su ambiente interno en una condición constante a pesar de los cambios en su ambiente externo.

hormone A chemical messenger that regulates the activities of cells in an organ or group of organs, or affects the activities of all the cells in the body.
hormona Un mensajero químico que regula las actividades de las células en un órgano o grupo de órganos, o afecta las actividades de todas las células en el cuerpo.

host An organism (animal or plant) that harbors (provides food and a place to stay) for another organism, such as a virus.
anfitrión An organismo (animal o planta) que hospeda (provee alimento y albergue) a otro organismo, como lo es un virus.

hygiene Things people do to stay healthy.
higiene Cosas que hacen las personas para mantenerse saludable.

hypothalamus A major gland of the endocrine system that is part of the brain.
hipotálamo Una glándula principal del sistema endocrino que es parte del cerebro.

I

immune system Body system that fights disease.
sistema inmunológico El sistema del cuerpo que combate las enfermedades.

immunity A condition of being able to resist an infectious disease.
inmunidad La condición de poder resistir una enfermedad contagiosa.

immunization A medical treatment that helps protect you from disease.
inmunización Un tratamiento médico que ayuda a protegerte de una enfermedad.

incubation period The amount of time someone may be infected with a disease before he or she shows symptoms. During this time, an infected person is contagious.
período de incubación La cantidad de veces que alguien puede ser infectado con la enfermedad antes que él o ella muestre los síntomas. Durante este tiempo, una persona infectada es contagiosa.

infection A growth of germs in your body.

infección El crecimiento de gérmenes en tu cuerpo.

infectious agent Something that can get inside your body, multiply, and cause disease.

agente infeccioso Algo que puede introducirse en tu cuerpo, multiplicarse y causar una enfermedad.

inflammation A reaction to invading microorganisms; signs of this include redness, heat, swelling, and pain.

inflamación Una reacción a microorganismos invasores; síntomas de estos incluyen enrojecimiento, calor, hinchazón y dolor.

inflammatory response The body's reaction to invading bacteria or viruses; includes the release of a chemical signal and increased blood flow.

respuesta inflamatoria La reacción del cuerpo a bacterias o virus invasores; incluye la liberación de signos químicos y aumento en el flujo sanguíneo.

initial carrier The first person in a group to get a contagious disease.

portador inicial La primera persona en un grupo en contraer una enfermedad contagiosa.

intravenous (IV) Into a vein; a liquid solution is given to a patient directly into a vein.

intravenoso En una vena; una solución líquida dada a un paciente directamente en una vena.

involuntary muscles Muscles that move without conscious control.

músculos involuntarios Músculos que se mueven sin un control consciente.

K

kidney One of two excretory organs that filter the blood and remove waste products.

riñón Uno de los dos órganos excretorios que filtran la sangre y remueven desperdicios.

L

leukocyte A white blood cell.

leucocito Glóbulo blanco.

lymph nodes Structures in the lymphatic system that filter lymph and trap microbes.

nódulos linfáticos Estructuras en el sistema linfático que filtran linfa y atrapan microbios.

lymph vessels Structures that carry lymphocytes throughout the body.

vasos linfáticos Estructuras que llevan a los linfocitos alrededor del cuerpo.

English & Spanish Glossary

lymphatic system Body system that collects fluid lost by the blood and returns it to the circulatory system. It also produces some types of white blood cells.

sistema linfático El sistema corporal que recoge los fluidos perdidos por la sangre y lo devuelve al sistema circulatorio. También produce algunos tipos de células blancas.

lymphocyte A blood cell that protects the body from invading microbes.

linfocito Las células de sangre que tienen como trabajo principal el proteger al cuerpo de microbios que lo pueden invadir.

M

magnifies Makes something look larger but does not actually enlarge the physical size of the object.

aumentar Hacer que algo parezca más grande pero realmente no agranda el tamaño físico del objeto.

marrow Soft tissue inside the bones that makes red blood cells and stores fat.

médula Tejido suave dentro de los huesos que produce células rojas y almacena grasa.

meninges The linings (membranes) that cover and protect the brain and the spinal cord.

meninges Los recubrimientos (membranas) que cubren y protegen al cerebro y la médula espinal.

meningitis A viral or bacterial disease that results in inflammation of the lining surrounding the brain.

meningitis Una enfermedad viral o bacterial que tiene como consecuencia inflamación en el recubrimiento que circunda el cerebro.

metabolism The combination of chemical reactions that takes place in an organism; food is converted into energy that the organism uses to carry out its life processes.

metabolismo La combinación de reacciones químicas que se llevan a cabo en un organismo; el alimento es convertido en energía que el organismo usa para llevar a cabo sus procesos de vida.

microbe (or microorganism) An organism that can be seen only through a microscope.

microbio (o microorganismo) Un organismo que puede ser visto solamente a través de un microscopio.

microbiologist A person who specializes in microbiology.

microbiólogo Una persona que se especializa en la microbiología.

microbiology The study of microorganisms.

microbiología El estudio de los microorganismos.

microscopic So small you can see it only through a microscope.

microscópico Algo tan pequeño que solamente puede ser visto a través de un microscopio.

mineral A type of substance found in food that is needed in small amounts. The mineral calcium is needed for strong bones and teeth, and the mineral iron is needed for healthy red blood cells.

mineral Un tipo de sustancia que se encuentra en los alimentos y que es necesaria en pequeñas cantidades. El mineral calcio es necesario para mantener huesos y dientes fuertes, y el mineral hierro es necesario para mantener células rojas saludables.

mitochondrion (plural, mitochondria) Provides the cell with energy.

mitocondria Le provee a la célula la energía.

model A way of representing something in the world to learn more about it.

modelo Una manera de representar algo en el mundo para aprender más sobre ello.

mucus Sticky, wet material in your nose and other organs.

moco Material mojado y pegajoso en tu nariz y otros órganos.

muscular system The body system, made up of the muscles, which moves the parts of the body.

sistema muscular El sistema corporal compuesto por los músculos, los cuales mueven las partes del cuerpo.

mutation A permanent change in the genetic material of a cell.

mutación Un cambio permanente en el material genético de una célula.

English & Spanish Glossary

N

nervous system Contains the brain, spinal cord, and an enormous number of nerves that carry messages about what is happening outside the body to the brain and messages from the brain to all parts of the body.

sistema nervioso Está compuesto del cerebro, la médula espinal y una cantidad enorme de nervios que llevan mensajes al cerebro acerca de lo que está sucediendo fuera del cuerpo y mensajes del cerebro a todas las partes del cuerpo.

noncommunicable disease A disease that cannot be passed on to other people by the person who is sick.

enfermedad no contagiosa Una enfermedad que no puede ser transmitida por la persona enferma a otras personas.

nucleus The control center of the cell.

núcleo El centro de control de la célula.

nutrients The useable substances in food.

nutrientes La parte utilizable de los alimentos.

O

organs Structures that have a specific function and are made up of different tissues.

órganos Estructuras que tienen una función específica y están hechas de diferentes tejidos.

organ systems Groups of organs that have related functions.

sistemas de órganos Grupos de órganos que tienen funciones relacionadas.

organelle A specialized structure in a cell.

organelo Una estructura especializada en una célula.

osteomyelitis An infectious bone disease caused by bacteria or fungi.

osteomielitis Una enfermedad del hueso contagiosa causada por bacteria u hongo.

outbreak Occurrence of more cases of a particular disease than expected in a given period of time or a given area.

epidemia La ocurrencia de más casos de los esperados de una enfermedad en particular en un período de tiempo o en una área.

ovaries Female reproductive organs that produce eggs.

ovarios Órganos reproductivos femeninos que producen huevos.

P

pandemic An epidemic that has spread throughout the world.

pandémico Una epidemia que se ha propagado a través del mundo.

parasite Organism that lives and feeds either inside of or attached to another organism and does harm to that organism.

parásito Organismo que vive y se alimenta tanto adentro de o pegado a otro organismo y causa daño a ese organismo.

pasteurized To heat food to a temperature that is high enough to kill most harmful bacteria.

pasterizado Calentar comida a una temperatura que es lo suficientemente alta como para matar la mayoría de las bacterias dañinas.

photosynthesis The process by which plants make sugar and oxygen, using light, water, and carbon dioxide.

fotosíntesis El proceso por el cual las plantas producen azúcar y oxígeno, usando luz, agua y bióxido de carbono.

Project Board A space for the class to keep track of progress while working on a project.

tablón de proyectos Un espacio para que la clase mantenga evidencia del progreso mientras trabaja en un proyecto.

prokaryote A single-celled organism that does not have a nucleus.

procariota Un organismo unicelular que no tiene núcleo.

protein A type of substance found in food that builds muscle and body tissue.

proteína Un tipo de sustancia encontrada en los alimentos que forma los músculos y el tejido corporal.

pulse The surge of blood in an artery as the blood is pumped by the heart.

pulso El aumento de sangre en una arteria mientras la sangre es bombeada por el corazón.

English & Spanish Glossary

pus A white, or slightly yellow or green substance that your body develops in response to an infection; it is made up of dead skin, white blood cells (that fight infection), and some bacteria.

pus Una sustancia blanca o levemente amarilla o verde que tu cuerpo desarrolla como respuesta a una infección; se compone de piel muerta, glóbulos blancos (que combaten la infección) y algunas bacterias.

Q

quarantine Keeping infected people away from healthy people.

cuarentena Mantener a las personas enfermas alejadas de las personas saludables.

R

ratio A comparison of two numbers.

razón Una comparación de dos números.

representation A likeness or image of something.

representación Un parecido o imagen de algo.

reproductive system The body system that is specialized to produce sex cells (eggs and sperm) and sex hormones.

sistema reproductivo Sistema corporal que se especializa en la producción de células sexuales (huevos y espermatozoides) y hormonas sexuales.

respiratory system An organ system that delivers oxygen to the blood and removes carbon dioxide, a waste gas, from the blood. It includes the nostrils, trachea, and lungs.

sistema respiratorio Un sistema de órganos que lleva oxígeno a la sangre y remueve bióxido de carbono, un gas de desperdicios, de ésta. Incluye las fosas nasales, la tráquea y los pulmones.

S

sanitation The disposal of sewage and waste.

sanidad La eliminación de aguas residuales y desperdicios.

sentinel case A person who begins the spread of disease in a group.

caso centinela Una persona que comienza la propagación de una enfermedad en un grupo.

simulate To imitate how something happens in a real-world situation by acting it out using a model.
simular Imitar cómo algo sucede en una situación de la vida real, representándola utilizando un modelo.

simulation The process or act of imitation or acting out.
simulación El proceso o acto de imitar.

skeletal system The internal framework of bones in the body that provides shape and support for the body, allows it to move, protects the organs, produces blood cells, and stores minerals and other materials.
sistema esqueletal El armazón interno de huesos en el cuerpo que provee forma y sostén para el cuerpo, le permite el movimiento, protege los órganos, produce las células sanguíneas, y almacena minerales y otros materiales.

spleen Filters blood by removing harmful bacteria and viruses, activates B-cells, and destroys old red blood cells.
bazo Filtra la sangre removiendo las bacterias y los virus dañinos, activa las células B, destruye los glóbulos rojos más viejos en la sangre.

spore In bacteria, a dormant structure that allows the bacterial cell to survive unfavorable conditions. In fungi, a cell that develops into a new organism.
espora Referente a bacterias, una estructura inactiva que permite a la célula bacterial sobrevivir en condiciones desfavorables. En los hongos, una célula que se desarrolla como un organismo nuevo.

sputum Matter that is coughed up and mixed with saliva.
esputo Materia que es tosida y mezclada con saliva.

stool Solid waste that is produced by the digestive system.
materia fecal Desperdicio sólido que es producido por el sistema digestivo.

sweat glands Structures in the skin that remove water and some salts.
glándulas sudoríparas Estructuras en la piel que remueve el agua y algunas sales.

symptom An indication in your body that you have a disease.
síntoma Una indicación en tu cuerpo de que tienes una enfermedad.

English & Spanish Glossary

T

testes (singular, testis) Male reproductive organs that produce sperm.

testículos Órganos reproductivos masculinos que producen esperma.

thymus An organ that is important in the development of a type of white blood cell.

timo Un órgano que es importante en el desarrollo de un tipo de glóbulos blancos.

Project-Based Inquiry Science

tissues Groups of cells that are similar in structure and function.

tejidos Grupos de células que son semejantes en estructura y función.

toxic Poisonous.

tóxico Venenoso.

toxin A poisonous substance.

toxina Una sustancia venenosa.

trachea (windpipe) A tube that carries air to the lungs.

tráquea Un tubo que lleva aire a los pulmones.

U

ureter One of two tubes that carries urine from each of the kidneys to the urinary bladder.

uréter Uno de dos tubos que llevan la orina de cada uno de los riñones a la vejiga urinaria.

urethra A tube through which urine flows from the body.

uretra Un tubo por el cual la orina fluye fuera del cuerpo.

urinary bladder An organ that stores urine until it is eliminated from the body.

vejiga urinaria Un órgano que almacena la orina hasta que es eliminada del cuerpo.

urine A watery fluid produced by the kidneys that contains wastes.

orina Un fluido líquido producido por los riñones que contiene desperdicios.

uterus (or womb) Hollow, muscular organ of the female reproductive system in which a fertilized egg develops.

útero (o vientre) Órgano muscular hueco del sistema reproductor femenino en el cual se desarrolla un huevo fertilizado

V

vaccination The process by which a substance that protects a person from a disease is given.

vacunación El proceso por el cual se le provee una sustancia a una persona para protegerla de una enfermedad.

vaccine A substance that protects a person from a disease; from the Latin vacca, for cow.

vacuna Una sustancia que protege a una persona de una enfermedad; del latín vacca que proviene de la palabra vaca.

vacuole A storage area for food and water.

vacuola Un área de almacenaje para alimento y agua.

English & Spanish Glossary

vas deferens A tube that carries sperm from the testes in males.

conducto deferente Órganos reproductivos masculinos que transportan esperma de los testículos.

vein A blood vessel that carries blood back to the heart.

vena Un vaso sanguíneo que lleva sangre de vuelta al corazón.

ventricles Muscular heart chambers that pump blood through the arteries.

ventrículos Cámaras musculares del corazón que bombean sangre a través de las arterias.

villi (singular, villus) Tiny, finger-like structures that protrude from the inside surface of the small intestines.

vellosidad Estructuras parecidas a dedos diminutos que sobresalen de la superficie interna de del intestino delgado.

vitamin A type of substance found in food that is needed in small amounts. It helps the body in many ways, such as keeping it from developing certain diseases.

vitamina Un tipo de sustancia que se encuentra en los alimentos y que es necesaria en pequeñas cantidades. Ayuda al cuerpo de muchas maneras, una de ellas es evitar que el cuerpo desarrolle ciertas enfermedades.

voluntary muscles Muscles that can be moved by conscious control.

músculos voluntarios Músculos que pueden ser movidos por un control consciente.

W

wasting syndrome A disease resulting from AIDS and affecting the hormones that metabolize food.

síndrome de desgaste Una enfermedad que resulta del SIDA y afecta las hormonas que metabolizan los alimentos.

white blood cells (or leukocytes) Blood cells whose main job is to protect the body from invading microbes.

glóbulos blancos o leucocitos Células de sangre cuya función principal es proteger al cuerpo de microbios invasores.

Index

Index

Index

Index

Project-Based Inquiry Science

Index

IT's ABOUT TIME®
HERFF JONES EDUCATION DIVISION

84 Business Park Drive, Armonk, NY 10504
Phone (914) 273-2233 Fax (914) 273-2227
www.its-about-time.com

Publishing Team

President	**Creative Director**	**Production/Studio Manager**
Tom Laster	John Nordland	Robert Schwalb
Director of Product Development	**Assistant Editors, Student Edition**	**Production**
Barbara Zahm, Ph.D.	Gail Foreman	Sean Campbell
Managing Editor	Susan Gibian	**Illustration**
Maureen Grassi	Nomi Schwartz	Dennis Falcon
Project Development Editor	**Assistant Editors, Teacher's Planning Guide**	**Technical Art/ Photo Research**
Ruta Demery	Danielle Bouchat-Friedman	Sean Campbell
Project Manager	Kelly Crowley	Michael Hortens
Sarah V. Gruber	Edward Denecke	Marie Killoran
Safety and Content Reviewers	Heide M. Doss	**Equipment Kit Developers**
Edward Robeck	Jake Gillis	Dana Turner
Barbara Speziale	Rhonda Gordon	Joseph DeMarco

Picture Credits

Page 5 — Fotolia/ photoaged
Page 14 — istockphoto.com/ Dra Schwartz
Page 17 — istockphoto.com/ Laurence Gough
Page 18 — istockphoto.com/ Laurence Gough
Page 23 — istockphoto.com/ Christine Gonsalves
Page 24 — istockphoto.com/ Robert Adrian Hillman
Page 25 — istock photo.com/ Joy Fera
Page 28 — John M. Patterson
Page 28 — micrographia.com
Page 31 — Custom Editorial Production Inc
Page 32 — It's About Time/ Bob Anderson
Page 32 — It's About Time/ Bob Anderson

Page 39 — Custom Medical Stock Photo
Page 40 — photodisc
Page 43 — Michael Johnson
Page 50 — Photo Insolite & V. Gremet/Science
Page 51 — Custom Medical Stock Photo
Page 55 — Department of Microbiology, Bioze
Page 59 — Robert Thom
Page 61 — Jaimie D. Travis
Page 63 — istockphoto.com/ Dagmara Ponkiewska
Page 64 — istockphoto/abdone
Page 85 — photodisc
Page 89 — istockphoto.com/ Yiannos
Page 90 — istockphoto.com/ Oleg Prikhodko

Page 96 — istockphoto.com/ Galina Barskaya
Page 99 — istockphoto/ Carl Subick
Page 102 — istockphoto/ Nancy Louie
Page 110 — istockphoto.com/ Nick Vernooy
Page 119 — istockphoto.com/ North Georgia Media LLC
Page 138 — istockphoto.com/ Sharon Dominick
Page 145 — Chris Hutchison
Page 149 — Fotolia/ Christos Georghiou
Pages 21, 56, 66 — Wikipedia/Creative Commons
All Illustrations — Dennis Falcon
Technical Art — Sean Campbell, Michael Hortens, Marie Killoran